NORTHAMPTON COMMUNITY
DISCARDED
COLLEGE LIBRARY

Mature Audiences

A volume in the series
Communications, Media, and Culture
George F. Custen, series editor

Mature Audiences

Television in the Lives of Elders

Karen E. Riggs

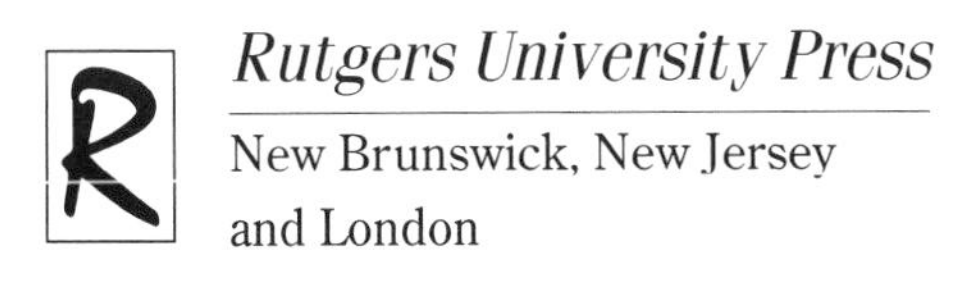

Rutgers University Press
New Brunswick, New Jersey
and London

Library of Congress Cataloging-in-Publication Data

Riggs, Karen E.
Mature audiences : television in the lives of elders / Karen E. Riggs.
p. c. — (Communications, media and culture)
Includes bibliographical references (p. 177) and index.
ISBN 0-8135-2539-X (cloth: alk. paper) — ISBN 0-8135-2540-3 (pbk.: alk. paper)
1. Television and the aged—United States. 2. Mass media and the aged—United States. 3. Aged—United States—Psychology. 4. Aged—Communication—United States. I. Title. II. Series.
HQ1064.U5R546 1998
306.4′85′0846—dc21 97-49866
CIP

British Cataloging-in-Publication data for this book is available from the British Library

Composition by Colophon Typesetting

Manufactured in the United States of America

For Gloria Howard,
Murder, She Wrote fan extraordinaire,
and Buck Howard,
who loved watching pro wrestling on Saturday afternoons

Contents

Preface

In her book *The Fountain of Age*, Betty Friedan remarks that "the dread of age has become so ingrained in our consciousness that even those who study or report on age seem to seek only the fountain of youth."[1] Friedan champions the rich possibilities of aging—an extension of life's personal journey—over the negatives portrayed in mass media and scholarship alike. Her accusation against those who report on or study the elderly has special resonance for the study of mass communication. Friedan observes that writers either avoid the unpleasant topic of old age entirely or characterize it as a problem. Three decades ago, anthropologist Margaret Clark similarly observed that few of her peers undertook aging studies, because most saw the subject as a "morbid preoccupation—an unhealthy concern, somewhat akin to necrophilia."[2] Years later, cultural anthropologist Barbara Myerhoff also noted that scholars avoid a subject that they find "vaguely repellent to them."[3] Clark and Myerhoff seemed to believe that ethnographers avoided aging studies because of their own fears about aging.

Like Friedan, Myerhoff, and Clark, I prefer not to perceive old age as a problem, the penultimate problem, just before death. Their research has been informed by a desire to study older people in a meaningful period of the life cycle. It is a period about which we, as scholars—in particular, those of us who study mass media—know relatively little.

This modest set of case studies is a means for me to add my own voice to those of these three women. Television is a significant phenomenon in the lives of the elderly, as social scientists have found, yet what has gone unexplored is how the phenomenon fits into older people's everyday lives. We know, for example, the uses older people make of television, but we do not know much about how those uses mesh with personal circumstances and social identities. This work is explores part of that terrain.

My own research interest in elderly people took thirty years to awaken. I have always spent much time in the company of the aged and have considered that to my benefit. I was surprised, however, to slowly realize that my impression of old people was narrowly informed.

One of my earliest memories was sitting in Great Aunt Sibbie's expansive lap every weekday afternoon in the Columbus, Georgia, public housing projects. After she would put the collard greens on to cook, she would rock me to sleep as she watched her "stories." I don't remember much about *As the World Turns*, but I recall Mr. Clean quite well. I used to fold my arms and wrinkle my forehead as I tried to re-create this advertising character from Aunt Sibbie's house for my mother and father. It was at about this time, when I was 3 or 4 years old, that I began to perceive television as a *natural* part of the lives of old people.

After Aunt Sibbie died and I was in school, I spent the majority of my free time around other elderly relatives, who looked after me while my parents worked. My paternal grandparents, who lived next door to us in a four-room, tin-roofed house with a cast-iron wood stove and a well pump in the backyard, did not have a television. They had a harmonica and a banjo. My maternal grandparents, who lived up the road, owned a set, but, except for *Lawrence Welk*, I don't remember seeing it turned on as long as they continued to work in the cotton mill, plow the garden, and travel the countryside for my grandfather's preaching duties. It was after my maternal grandparents retired that the television set started pouring out its programs to them, not just in the evenings, as I was accustomed to in my parents' home, but in the afternoons as well.

Pretty soon television was their only leisure. My grandparents weren't able to plow anymore, and my grandfather had run out of fire and brimstone. They became loyal fans of the CBS soaps, Merv Griffin's game shows, and *Donahue*; these elderly folk, who had never used the word "pregnant" in front of me, were now watching Nikki hop into bed with Jack one day, Victor the next. They never missed the evening news and particularly looked forward to "the local." At night, *The Dukes of Hazzard* would race across the screen in their den, and they could barely contain themselves while waiting to find out who shot J.R.

Their mornings were still tubeless for many years, until the family grew too nervous about my grandfather's venturing off to McDonald's in the car and forgetting where he was. When he finally surrendered the car keys, the television became his only link, other than the family, to the world outside. My grandmother's interest in the medium increased, too, but she continued to read every word of the evening paper and listen to Larry King on the radio at night in bed. A relative gave them a VCR just as my grandfather's Alzheimer's disease started to advance. Soon, he was hooked on a home video someone had provided of the "sacred harp singing" at his boyhood country church. My grandmother had to restart it for him several times a day. Similar things were happening in the lives of their neighbors, other blue-collar white couples who had moved from the farm back in the 1940s and settled into their now decaying neighborhood when it was new.

After college, when I moved to a south Florida resort town, I was shocked when my image of worn-out old people, clinging to the television set for life, was contradicted by the people I met. As a newspaper feature writer, I began to meet well-off old people, who—I could barely believe it—had been to college in the years when my grandparents were weaving at a loom in the mill. Energetic and informed, these retirees from the North played tennis and golf and sported tans on their relatively wrinkle-free faces, arms, and legs. They all got the *Wall Street Journal* delivered, and their neighbors were people with names like du Pont and Ford and Regan. They were at the club during *Oprah!* and never knew she was on.

Later, in graduate school, as I began to ponder questions of how media work in the everyday lives of people, I returned to these disparate images of the worn-out and well-off elderly. The more I would read in the social science literature that indicated a homogeneous retirement experience among older Americans, the more eager I was to examine diverse audience environments for myself. When someone suggested I look at a local retirement community, I jumped at the chance. These, after all, would be people I knew—the elderly, as many of the people I've spent my life around have been, and part of the educated middle class, which I had come to consider myself a member of. It turns out, actually, the people in that community, like all the other elderly people I've met in connection with my research, were brand new to me.

Acknowledgments

In addition to the dozens of elders who have made this account possible, I must acknowledge the invaluable professional and collegial assistance I received. Institutional support included a grant from the University of Wisconsin–Milwaukee Graduate School and a fellowship from the university's Center for Twentieth Century Studies. The center's director, Kathleen Woodward; staff; and visiting fellows provided enormous intellectual support. I am indebted to students, colleagues, and advisers who also took an interest. Among the students who participated in helping me gather information and transcribe interview tapes were Alison Farmer, my research assistant, and other graduate students: Bob Pondillo (who helped produce an earlier version of chapter 4 as a conference paper), Jill Gilbertson, Rebecca Leichtfuss, Jill Gilbertson, Ai-Ling Huang, and Shan-Wei Sun. Shaun Navis and Luz Gonzalez helped as undergraduates. Colleagues in the Department of Mass Communication at the University of Wisconsin–Milwaukee lent me immeasurable and unwavering support; I especially appreciate the interest and advice of Chair David Pritchard. Maria Flores and other office staff members helped at many stages. My former professors at Indiana University read and advised me on drafts of papers in which ideas were developed that eventually became part of the manuscript for this book. I am indebted to Kathy Krendl, Christopher Anderson, Carol Greenhouse, Michael Jackson, Louise Benjamin, and especially my dissertation adviser and friend, Ellen Seiter.

Ellen not only labored through the dissertation portion of this research but enthused and suggested improvements as the work grew. I also owe gratitude to other professors with whom I worked at Indiana: Michael Curtin, Susan Eastman, Bonnie Kendall, and Jim Potter. I must recognize the early support I received from Mike Budd, Fred Fejes, and Clay Steinman, all of whom advised me at Florida Atlantic University as a master's student; without their encouragement, I would never have pursued research. Among other scholars in the field, Ted Glasser has provided generous critiques of portions of the manuscript at various stages, and Veronique Nguyen-Duy has also helped with her comments.

My family has been extraordinarily supportive, including my parents and grandparents, two of whom (my father and grandfather) died as this book was being developed. My children, Austin and Emily Riggs, both learned to walk and talk and have entered school as various branches of this research have been undertaken; they have cheered me on with unbridled enthusiasm. Tom Riggs has been a tireless editor and patient friend, and his many contributions can never be sufficiently acknowledged.

Leslie Mitchner, editor in chief at Rutgers University Press, envisioned a broader and much better book than I initially had. As I have produced it, she has provided brilliant analysis of too frequently not-brilliant pages and has helped raise the bar that I was to hurtle until I could comfortably take pride in this modest effort. My thanks also to series editor George Custen, who welcomed the book even as he ably found places where it needed improvement in the final stages of writing. Marilyn Campbell and Grace Buonocore made invaluable improvements to the manuscript.

Mature Audiences

One

Television and the Elderly Audience

In late 1996, a new television commercial aimed at investors appeared, featuring a couple of aging baby boomers. The commercial came off as clever because it defied common knowledge about age. A wife and husband were talking to each other on the telephone about an unnamed male member of the family, who had capriciously just gone off to Europe to study French impressionism after quitting his excursion into photography. "Why can't he grow up?" the husband asks, leaving the viewer to imagine a 20-something son as the transgressor. "He's *your* father," the wife answers. It was supposed to be funny because older people were not *supposed* to behave like 20-somethings.[1]

Cultural stereotypes about the elderly depict a passive group that is either too tired or too inconsequential to engage in meaningful acts. This book insists that such a portrait is overly general and misses the complexities of late life as it is defined in the lives of Americans, particularly through their involvement with media. It shows how activity and participation differ qualitatively as well as by degrees among older people living in the United States. The patchwork of stories that conjoin here do not uniformly celebrate elders' strength but probe some of the multifarious ways in which they negotiate the constraints against them. Ultimately, the book proposes to help define the work of older Americans to tailor their major leisure activity—watching television—for the lives they have been able to fashion for themselves at the opening of the twenty-first century.

A Diverse and Creative Audience

I discern flickers of hope for the future through the creative ways in which older Americans are using television today. Even the most stereotypical, tradition-bound uses that older people make of television are not always what they seem to be. Creativity may exist in almost any audience role, and I have found older viewers to be very much in control of their viewing choices and the purposes they are trying to serve with these. This book presents four examples of older viewers using television in ways that help them relate their public concerns with their everyday, private lives.

Chapter 2 takes a look at a traditional use of television by older women and poses it as a creative, constructive act. Women of diverse backgrounds, all fans of the television mystery show *Murder, She Wrote*, talk about why they find this program, long dismissed as dull and unimportant by television critics, so compelling. The study suggests that older people, women particularly, actively looked to this show to revalidate their lives in the midst of shifting life circumstances. These viewers ritually engaged the hour-long episodes of the television mystery drama as a method of galvanizing their convictions about the world and assuring themselves of their place in it. This chapter provides a textual analysis of *Murder, She Wrote*, along with another classic mystery drama, *Perry Mason*. The chapter relies on interviews to show how different women perceive *Murder, She Wrote* and incorporate it into their viewing routines.

Chapter 3 describes what may be the germ of public discussion encouraged by new texts of television. At Woodglen, a retirement community in the American Midwest, people are using television in ways that are different from what we have come to expect from the aged. Their use of the medium, in fact, seems to have more to do with a desire to exercise power over their own lives than with classic motivations of companionship, escapism, and security.[2] Through their reliance on and manipulation of nontraditional television channels, these viewers can be seen as pioneers. They are prototypical viewers of new nonfiction television. The uses they are making of such programming ultimately will help shape the conventions of these forms and influence television's changing relationships with its viewers.

This chapter describes how a group of elderly Americans are reacting to what they have found to be premium content of public and cable television channels. The nonprofit Cable Satellite Public Access Network, C-SPAN, and its clone, C-SPAN 2, offer much promise for diversity of viewpoints in the future. The truth is now, however, that both channels proceed from an orientation that perceives U.S. issues in terms of a two-party, Washington-focused, representational system of politics. Telephone callers to the channels, who converse on air with a host, largely reflect this orientation. The calls are not screened, however, so, occasionally, less traditional views are aired. It may be possible that less traditional thinkers will become attracted to these channels and other outlets like them. The appearance of such channels as C-SPAN on cable systems suggests a reconstitution of the role of the audience that may produce more activity in the home and, possibly, more activism. As more people use the new channels and the new, more powerful technologies being developed, public communication via the mass media could be infused with new political meaning.

Drawn from ethnographic methods, chapter 3 introduces the Woodglen community and situates it amid the phenomenon of the American retirement community. It describes how residents negotiate their domestic and communal spheres to construct meaning from television texts, with emphasis on achievement of public participation through "premium" nonfiction programs. The chapter explains cultural factors that lead residents to prefer particular programs over others within their most favored genre, news; it shows how a community logic of television reflects hierarchies of taste that are constructed through class experience.

Chapter 4 illuminates another context in which elderly people are using television to address the public arena, this time through producing their own public-access television shows or appearing on the programs of their contemporaries. Public-access television, which is still in its relatively early stages of development through cable, offers communities talk at the grass-roots level, outside the local power structure. In order to gain access to such channels, organizations have only to provide the necessary capital for modest productions. At this point, the downside of public access is that relatively few people watch it, and fewer use it to distribute information. A chicken-or-egg issue exists as well: Will the telecast of public meetings on such channels detract from citizen attendance of them or ultimately spur interest in local government and civic affairs? This is a question that is analogous to the infancy of broadcast media: producers of live events, such as sporting contests, resisted the idea of having broadcast coverage, because they feared it would turn ticket holders away; the opposite turned out to be true. The impact of public access seems marginal at present. Nevertheless, public access is growing, a trend that may signal a broader interest in such communication channels in the years to come.

In addition to three access channels for institutional use, the city of Milwaukee's telecommunications authority sponsors two local-access channels. These two channels are for use by the citizenry, with or without the involvement of clubs and organizations. These channels, in use since 1986, offer private parties the opportunity to train in the use of production equipment, borrow the equipment, produce original programs, and distribute them locally on cable television. Older citizens' involvement with these channels has provided a significant component of their telecasts. Among elderly "producers," religious programs have been popular, and so have personalized productions such as one octogenarian's autobiographical video. As cable-access channels continue to proliferate and attract attention, who is to say which voices may take advantage of such free production opportunities? Chapter 4 analyzes production efforts by five African Americans who have each attempted to inject their moral concerns into the public sphere with a style of activism not normally associated with the elderly.

Chapter 5 does not examine the television experience of a homogeneous group but considers members of some of the most overlooked categories of an already overlooked age group. It scrutinizes the television involvement of a variety of elders from minority and immigrant communities. Among the voices heard from in this chapter are American Indians and immigrants from Russia and Laos, but also heard from are gay and lesbian elders who have lived their entire lives in the United States as an oppressed other. The diverse uses to which these individuals have put television have much to do with staking out their personal identities and with making their way in their social worlds.

The Fictive "Elderly Audience"

This book presents a logical progression from the research that British cultural scholar John Tulloch did in his parents' retirement community in Bournemouth, England, in the late 1980s. Tulloch has been one of a handful of scholars to look at older audiences since the mid-1980s, and his approach highlights not their commonalities but the differences he found among the residents of this homogeneous village. Tulloch was able to identify a variety of personal characteristics and lifestyle factors that contributed to how older people fit television into their lives. In the tradition of British audience studies, he especially highlighted gender and class distinctions, and he considered the social context of the individual a major problematic. For example, Tulloch relayed an interview with a couple who said they enjoyed a medical soap opera together, and in addition to amplifying their experience as an outgrowth of gender and class, Tulloch went on explain that the man liked the show because it sometimes included the topic of diabetes, a disease from which he suffered, and the woman liked it because its themes often paralleled her experience as caregiver for the ill husband.[3]

Positing the audience as a collection of idiosyncratic positions, Tulloch did not work to establish the Bournemouth village as a cohesive cultural group—a theme that I have suggested among the people whose stories are told in chapter 3—and he does not establish the villagers as representative of any sort of broader British or Western experience. He successfully positions the audience as fictive, in the tradition of Ien Ang, John Hartley, David Morley, and Janice Radway.[4] He shows that researchers must talk about the various experiences and categories that pull on us to form our very subjectivities in ways that allow our media experiences to develop with certain peculiar regularities, ones that we cannot extend to any broad public.

The elders whose stories this book tells point to a number of observations about this most pervasive medium. For one, they defy the stereotype that older people are monolithically unimaginative in their approach to television. Instead, these people make explicit connections between the television content they use and other forms of communication. For them, television texts are not discrete objects. These viewers make fluid choices about channels and content according to their needs at any one time. They consider the text of one program when they evaluate that of another. They talk intertextually about content they have collected and analyzed from the texts of print, television, radio, and film, just as we have come to expect that members of the so-called baby boomer and X generations also incorporate what they come to know through use of the videocassette recorder, the Internet, and telephone technologies in their sweep of media texts. For elderly Americans today, contrary to what we may have imagined, almost everything that is culturally relevant is mediated. Instead of remaining out of touch with the messages of technology, these elderly engage in talk with one another and with family members that is filled with influences from media. They may reject or avoid information that does not resonate with their own experiences, but they nimbly and aggressively seek media reference points that they find useful in their everyday lives. Younger people might find these viewers' television habits dull, but, for these elders, the act of assembling the content they will use can be both exciting and rewarding.

Second, as these viewers die and baby boomers replace them in old age, the new generation of older audience members will express a stronger will to control the technology in their homes. Although we typically see in American culture a tendency for people to grow more conservative in their habits as they age, it is impossible to say what will occur in the television viewing audience as the first generation to grow up with television—and which readily adopted its related technologies—begins to manipulate the television apparatus comfortably in their leisure years. It is safe to say that, at the very least, these baby boomer retirees will have VCRs that do not perpetually flash twelve o'clock. Having long felt comfortable with the intricacies of the technology in their homes, I expect, they will find themselves feeling less bound to the particular kinds of content—conventional news shows and quiz shows, for example—that have been favored by the elderly cohorts ahead of them.

These baby boomer elders will come of age, so to speak, in a time of drastic media convergence. We cannot predict the state of the mass media in twenty to thirty years. However, it is certain that the television, the telephone, and the microcomputer will become increasingly difficult to differentiate, and the generation of Americans who are now middle-aged will be accustomed to the technological demands and amenities suggested by advanced media systems by the time they reach old age. They will meet these demands and amenities in their jobs, in their leisure time, and through their children's experiences with them. It is significant, then, that the discomforts and experimentations experienced by the members of the elderly cohorts described in this book are uniquely constrained by historical circumstances. The people I met in connection with this book did not grow up taking electronic technologies for granted. By the time electronic media convergence began in earnest, their generation had already reached old age, not known as a time of ready acceptance of swift changes. (This is an ironic notion about elderly people, really, considering that most of them almost certainly face a compression of swift changes, such as the loss of a longtime partner, loss of employment, change in home arrangements, and changing power relations with children.)

Television's Ageism: Embedded in Everyday Culture

I have long felt that the ugliest aspect of television's treatment of older people was its arrogance: television's creative workers apparently recognized that elders lacked cultural—or market—significance, and they exploited that lack in their renditions of this broadly conceived demographic group. A signature incident of such maltreatment came on an episode of the successful, long-running NBC situation comedy *Seinfeld* during the mid-1990s. Jerry's parents, retirees visiting from Florida, had just informed him that they did not care for the parents of his friend George. They astonished Jerry by telling him that they found the Costanzas rude and annoying. "Hadn't you noticed?" they inquired of Jerry. "Well, I've noticed, but they're from *your* age group," he answered. "I didn't think you could detect abnormal behavior from your own kind."

Television's texts richly illustrate the capacity of popular media images—and the lack of them—to paint the elderly as an expendable component of society. In much of entertainment television, as in *Seinfeld*, elders have been represented on a token basis, as buffoons for comic relief, and as victims.[5] Rarely in television's history have programs highlighted older characters as vital individuals. The list of unproductive stereotyping is voluminous, especially in comedies. It includes such characters as Walter Brennan's daft Grandpa McCoy (*The Real McCoys*); Irene Ryan's superstitious, manipulative Granny Clampett (*The Beverly Hillbillies*); Redd Foxx's selfish, irrational Fred Sanford (*Sanford and Son*); Abe Vigoda's perpetually constipated Detective Fish (*Barney Miller*); and Alice Ghostley's oversexed, confused Bernice (*Designing Women*). Jim Backus's undiscerning Mr. Magoo, a cartoon character, and Noah Beery's countrified Rocky (*The Rockford Files*) illustrate the tendency for these comic images to spill over into other genres, in these cases, children's programming and prime-time drama. Soap operas often have had a benevolent matriarch (*Dallas*'s Miss Ellie) or patriarch (*One Life to Live*'s Victor Lord) or, occasionally, a more multidimensional figure, such as *All My Children*'s Phoebe Tyler Wallingford. Often, however, elders on such serials are portrayed sparingly or represent an evil threat to youthful patriarchy. A memorable example of the latter was Jane Wyman's Angela Channing (*Falcon Crest*), a ruthless, witchlike woman whose character embodied the corruption of inherited wealth and power. She presented a constant threat to the achievements of her hardworking, middle-aged relative Chase Giaberti.

Most often, though, elders are merely absent from television. NBC's "Must See TV"—prime-time Thursdays—has overwhelmingly ignored older figures or at least made them the butt of humor. And although they are a substantial component of the news audience, elders receive scant attention on these programs. Much of the network news that mentions older people frames them as victims or dependents through stories about such topics as health or economic worries. The aged have only recently begun to figure somewhat seriously in advertising, where they have finally been identified as a "market niche" that can be hailed in the name of products and services associated with an "active lifestyle," but, within the broader field of advertising, images of smart older consumers remain few.

Such pervasive treatment in mass media makes exceptions starkly noticeable, as in the cases of *Lears*, a glossy magazine for older women; *Driving Miss Daisy*, a sensitive film about growing old; and the richly developed character of Ruth Anne, the wise septuagenarian depicted in the quirky CBS melodrama *Northern Exposure*. The accrual of stereotypical representations, balanced with the general rarity of elders' images, speaks to our own cultural fears about growing old. We will be helpless, foolish, sick. We will be in the way. Worse, we will be invisible. The media, especially television, take on the hue of our own cultural knowledge about the misery of aging and instruct us to treat our own elders shamefully. The message is a spin on the old saying about children; this is a population who would do well to be neither seen nor heard.

To appreciate the paucity of reasonable representations of the aged in U.S. media, it is useful to consider the cultural specificity of these representations. In some traditional societies, the aged are prized and protected, appreciated for the sum of their hard work and enjoyed for their capacity to pass on the lore of the culture. Cultural scholars Jenny Hockey and Allison James have noted, for example, that among the Tallensi people the eldest man of the family remains its head, and, although he may no longer participate at the same levels in some ways, he retains a supervisory role.[6] In Western cultures, as Kathleen Woodward has pointed out, "the dominant trope for aging has been the decay and decline of the body."[7] In some cultures, where age is viewed as a source of strength and wisdom rather than as a burden, aging gets greater popular attention in more pleasant ways. For example, in China, where the youngest national leaders are in their 60s and no strong cultural bias against the aged exists, television portrays the aged as vigorous, honorable citizens. So, too, in India, television representations of the elderly often convey a sense of respect for the wisdom and leadership status of the old—at least for men, demonstrating how age, like other social categories, ineluctably intersects with gender. (Indian popular culture often venerates older women for their value to home life.) As nations tend to become more Westernized, stereotypical representations of old age may develop, as they have in Japan. Once a very traditional society that honored its aged, Japan now wrestles with the problems of contemporary life in an industrialized society. A popular comic conflict portrayed on Japanese television is the modern wife's problem of coexisting alongside her mother-in-law in the home, as tradition may have dictated but feminist consciousness now rejects. The old mother-in-law, for better or for worse, gets stereotyped.

In short, Western stereotypical images of elders sell this group short. This book counters that set of images. It probes the meaning of television in the everyday lives of diverse older people in order to learn about both mass media and aging. To know more about the elderly is to better respect old age.

Developing Technologies and Changing Demographics

How different populations of elders will come to use such changing television channels and contents is unknown. The question then lingers, if retirees are dealing with a rapidly changing technological environment and their experiences surely are different from those that lie ahead for elders of different cohorts, why concern ourselves with them?

Understanding how "pioneer" audiences use particular technologies gives an important historical context. If a segment of society that essentially has been labeled "uncreative" and "static" is doing experimental things with media, this signals a creative capacity on the part of us all. If we can describe how such media users adapt to new resources for communication, as in the examples of the retirement community and the public-access program producers, we can better envision developments between media technologies and the oncoming parade of cohorts who will take their turns in the role of "elderly audience."

Elderly television audiences will burgeon. While elders may tend to have less disposable income than younger people, income is not the only indicator of wealth—and of consumer status. The U.S. Bureau of the Census has reported that, of American families whose members are 55 or older and retired, their mean net worth is two hundred thousand dollars, substantially more than averages for the so-called baby boomer groups.[8] As life expectancies increase and the health care industry maneuvers to help people lead more "normal" lives in old age, the so-called gray market will grow in significance for commercial media.[9] Consider the current shift in the U.S. population: in 1993, 22 percent of Americans were 55 or older; by 2010, that proportion will increase to 25 percent.[10] The advertising industry market model that has been associated with a pursuit for younger demographics in the hope of greater sales is now in question. Four particular developments make the study of elderly viewers increasingly important in the context of the state of the television industry:

1. Segmentation of the audience due to proliferation of channels. Market segmentation has sent networks scurrying for new niches and may spur reconsideration of faithful stereotypes of the elderly.
2. The nation's demographic shift, signaling not only longer lives for the current crop of older viewers but a displacement of the familiar curve of audience age groups as the formidable generation of baby boomers slips into retirement and an extended old age. The boomers' high disposable income replacements will number a much smaller group.
3. The burgeoning health care market has found a "natural" link with senior consumers. Ads that "market" active elder lifestyles and assisted living products have become common on television.
4. While media representations and everyday cultural relations may reflect a lack of respect of elders, this demographic group's political clout is silently growing. The American Association of Retired Persons is one of the largest lobbies in the country, and elders, who are growing in num-

ber, are among the most faithful groups of voters. Most Americans haven't heard much from the AARP, but the organization has been vocal in recent years on such issues as health care reform. Older baby boomers are now eligible to join AARP (at age 50), so this organization may improve its penetration of the political fog as time goes on. So far, AARP, with all its lobbying cachet, has not been able to make much progress in dispelling the cultural myth of elder as victim and dependent. For one thing, the news media all but ignore it.[11]

Elders and Interpretive Strategies

Arguing for the relevance of elderly populations as audiences for new technologies still may seem incongruous: if younger people are known to be more adaptive embracers of technological change, why pursue knowledge of the elderly on such a terrain? An answer can readily be found in audience studies: researchers here tend to believe that the most provocative observation that can be made from a technology is understanding how people make meaning from it through the cultural labyrinth in which they find themselves. Some of the most interesting media audience studies have centered on groups holding little official cultural power, yet they were perceived to play a role in the social significance of a particular technology.[12]

When we study the potential of a technology, its application to youth is initially most seductive. As researchers, and as a society as a whole, we tend to overlook the serious work that elders may bring to media consumption. I have found, for example, that some of them look upon their use of television texts as a vocation in the absence of employment in their retirement years. Watching television, quite plainly, is serious business to many elderly people. When many of us think of old people watching television, we envision such shows as *Wheel of Fortune*. Elder fans of such programs have told me one after another that they like the way these shows keep their minds active. But the quick reaction from younger people toward these programs is one of derision. It is easy to ridicule our elders for whiling away their evenings with "pointless" game shows, but let's look around us. Practically every social group has provided fan cultures for devalued texts, from *Sports Center* to *Days of Our Lives*. Many people similarly ridicule the substance of such programs as *Barney* and *Mister Rogers' Neighborhood*, but advocates of public television point readily to the usefulness of a purple dinosaur and a sweet-talking old man who prepare children to use their emerging intellect and social capacity. As for *Wheel of Fortune* and such programs, can we not similarly see the usefulness of television content that feeds the intellectual resilience and sense of connectedness to culture that exists for the aged?

We should knock down old stereotypes about social categories and understand how and why people use television in distinctive ways. In doing so, we can learn something about how older viewers carve out special meanings from television texts that are prepared largely without their concerns in mind. The corporate model of television programming traditionally has held older viewers practically in contempt, assuming their loyalty. The time has come for the system to adjust to the reality that the aged constitute a complex, creative audience that may be wilier than it was thought to be.

Old age is a ridiculously essentialized label. The television industry, for example, categorizes the older "demographic" as all Americans 50 and above. Scholars do studies on aging drawn from "samples" of populations 55, 60, and sometimes 65 and better. These sampling frames often reflect cultural uncertainty about who is old: Is it the one who is eligible for early retirement, for Social Security? Is it the empty nester, the one whose hair has turned gray? Is it any of these people, except for me? Is the standard different for women and for men, who have different life expectancies and, often, different stages at which they can begin collecting Social Security? Now that the eldest baby boomers are eligible to join AARP and some are touting "victories" over menopause or the corporate rat race, such groupings as the television industry's definition of old age are understandably in question. As this book demonstrates, people "50 and over" often do have many things in common, as the industry suspects, but it is misleading and unfair to generalize the life experiences of a 50-year-old to those of his or her 90-year-old grandparent. Gerontologists have found it useful at least to distinguish between life stages—the young old (60–75), the middle or old old (75–85), and the very old (85 and better).[13] It has become common for people as late as their middle 60s to identify with middle age, because they can expect to live another twenty years or so, they are still working for pay, and they may be involved in both financing their children's education and routine care for their own elderly parents or in-laws. It's hard to feel "old" when you are in relatively good health, you're still contributing to a retirement plan, and you make weekly trips to furnish your mother with prescription drugs and Depends. Chronological age has become a relatively unreliable means of sorting out the life course.

For the practical purposes of this book, I have interviewed people 60 and over.[14] I have resisted dividing them into neat ranges by years. Because some of the people seemed to have much more in common with individuals outside their "age group" than with others their own ages, it made sense to talk about old age in terms of life circumstances. I wanted to bring recognition to people's social ages, which are inflected by differing circumstances. I have treated old age as a complex field that cannot simply be unriddled by statements like "Sixty-year-olds say this but 70-year-olds say that." Instead, I have talked about people's experiences in such terms as widowhood, living arrangements, and the commitment of their time. Similarly, I have resisted the labeling of these people as "old" or as "seniors." (Many elders find this latter term extremely silly and demeaning.) I'm not sure that most people 60 and better will appreciate being identified as elders, but it was term with the broadest potential meaning and less disrespect that I could locate.

Gender, race, class, educational experience, religious identification, and regional differences come together to shape the identities of older Americans. These people cannot be perceived as a homogeneous group any more than any other age group can. In fact, different cohorts of elderly are distinguished by their disparate histories with media. Some, for example, were well into their adult lives before radio receded in prominence as television became dominant. Others were much younger when this occurred and saw the changeover from different perspectives. These differences mark elders' understanding of television in ways that are similar to baby boomers' experience of never knowing a life without television.

To try to relate the stories of a diverse assembly of real people, I have relied on qualitative methods, chiefly interviewing and observation in some attenuated media studies version of the ethnographic tradition. I have supplemented with such means as textual analysis when it has been important to trace connections between these elders and the television programs that have been so important in their lives. I have also relied on focus groups. (Although some scholars might be surprised to see focus groups linked with the more traditional ethnographic tools, I have found them to offer a complementary perspective; I have learned much through group talk that I believe I would never have learned by mere individual interviewing.) As laid out by Ellen Seiter et al. and others, ethnographic methods are a means to meeting the challenge of describing patterns without reducing people to essentialized categories. They help to establish the television consumer's role as "much more complicated than the poles of activity and passivity can accommodate."[15]

The people whose words appear in this book give us a glimmer of understanding of the creativity that people use in order to construct meanings that they find useful in their everyday lives. Typically, the successes they are looking for in this creative use of television are not very far from some version of the so-called American dream—acceptance by or admiration of peers, place in the community, cheating mortality, assurance that one's way of seeing the world is correct. What we do not yet know enough about is how changing mass communication systems can help elders construct themselves as citizens. It's up to media scholars, industry activists, and advocates for the elderly to foster ways for the aged to participate in public action. This is a territory about which we know too little. As Elihu Katz has pointed out, scholars need to probe "the incidence and conditions for conversation about public affairs"; in doing so, we can "explore the missing links between opinion and action."[16] In chapter 6, I will return to the subject of what television means in a broad sense for American elders, considering how their experiences might help shape the discussion of television's shifting meanings in an aging society.

Two

The Case of the Mysterious Ritual

What CBS has done—and this is absolutely true—is they've wrecked a wonderful kind of family gathering. A lot of family life, strangely enough, was centered around watching *60 Minutes* and then *Murder, She Wrote* on Sunday evenings. What people are telling us more and more—and this sounds like a con, but I do believe it—is that they're taping the new shows when they run on Thursdays, then watching them with their kids on Sunday nights.
—Angela Lansbury

When Angela Lansbury spoke these words in an interview with *TV Guide* (November 5, 1995), she did so with the knowledge that, for twenty-five years, *60 Minutes* had, for a cluster of older cohorts of American television viewers, marked the end of the week and that her own show had played a part in that ritual. In fact, if *Murder, She Wrote* had instead been the game of Clue, we could say that Jessica Fletcher did it in the TV room with a skewer—every Sunday night for eleven years. Older women, who like mysteries as a rule, ensured the show's perennial popularity.[1] Programs such as *Murder, She Wrote* and its early television cousin, *Perry Mason*, draw viewers in this demographic subgroup to a television form that rewards ritualized consumption—deliberate, routine uses that allow them to recover a certain deep structure of feeling.[2]

Women who once marked off their lives with the ceremonies of home-making and church or civic activities may sense a vertigo when these self-defining customs disappear.[3] They need new rituals to replace them, or they may come to rely more on other, existing rituals to seek those rewards of self-definition and a sense of order or progression. Many women—as well as some men—satisfy the need for some of these experiences through their viewing experience with mystery programs.[4] Absorbing oneself in a formula mystery—a book or a television program—allows one to dwell periodically in that liminal state between everyday reality and the familiar ground that is mystery's formulaic nonreality.[5]

While it is apparent that people engage mystery texts ritually, we can look to these texts to understand how their producers have made them suitable for such consumption. Typically, when textual critics examine television programs, they judge them on the basis of aesthetic qualities or their relevance to cultural relations. If a text cannot be judged aesthetically superior (*Hill Street Blues*, *Northern Exposure*, *Twin Peaks*) or be proposed as a site for cultural resistance (soap operas, *Cagney and Lacey*), critics likely will dismiss it as unremarkable. An appealing alternative is to see television viewing as ritual.[6] Textual and cultural critics who might approach the television mystery dismissively omit an important perspective. The classical mystery may not hold any surprises in form, but that is exactly the point. The people who flock to it do so out of the rewards it holds for them: its pleasant predictability, its preoccupation with life's gravest themes set against innocuous backgrounds, its concern with moral order and interpersonal conflict, and its assurance that chaos will be restitched into harmony. Programs such as *Murder, She Wrote*, whose structures are tightly formulaic, may cause us to notice their ritual nature because of their lack of innovation.

Such formula stories reflect culture as well as moralize about it. Faithful audiences enjoy this opportunity to confront problems that are relevant to their lives (in safely exaggerated settings) and solutions that ensure the triumph of order while giving rein to explore moral boundaries. (Although the subject of murder may not be so relevant for most viewers, the problematic surrounding the murder often *is*—motivations of the human psyche, troubled interpersonal relationships, as Jane Feuer has suggested about melodrama.)[7] Exploration of moral boundaries is a strong subtext of the murder mystery. Perry Mason, for example, virtually always won his case, ensuring that no murderer, no matter the motive, would get away without facing the authorities, but the murderer's motive was explicated and the murder act itself was framed as a rational, if usually immoral, act.

Perry Mason and *Murder, She Wrote* may seem to be pure escapist forms, but the connection between them and their audiences involves preservation of the self. For elderly people, the reassuring mystery presents a means to validate the self at a stage of life when one's identity is threatened in many ways by society as a whole. This preference for the genre is akin to the tendency for some elders to repeat their own stories, tales that measure out their lives and reward the teller with reminders of individual stability through time.[8]

Mystery and detective programs have supplied the broadcast networks with popular content fairly consistently since the days of radio. For example, *Perry Mason* was a radio serial drama on the CBS network from 1943 to 1955 before becoming a CBS television episodic drama, running from 1957 to 1966.[9] The mystery-detective genre peaked in popularity during the 1974–1975 season, when four series made the list of the top twenty rated programs by the A. C. Neilsen Company.[10] In the mid-1980s, CBS scheduled what would be its most successful mystery series ever, *Murder, She Wrote*, starring Angela Lansbury as aging amateur sleuth Jessica Fletcher.[11]

As the mystery genre receded from prominence, *Murder, She Wrote* served as a notable exception. Despite its undesirable skew, CBS was able to depend on it to deliver audiences for advertisers while much of the network's prime-time schedule grew weak.[12] Angela Lansbury herself has spoken out against what she considers a double standard in ratings logic. Likewise, the Gray Panthers, an advocacy group for older Americans, criticized the advertising industry's "ageism" and questioned its logic in light of national trends indicating the general aging of the U.S. population and the rising wealth of elder Americans.[13]

The classical mystery format has become tired and less reliable for networks, which compete with more channels and other entertainment forms for the attention of high-spending audiences. Advertisers desire a younger audience than the older women who flocked to *Murder, She Wrote* each Sunday evening. Near the show's termination, *Murder*'s competitor on ABC, *Lois and Clark*, drew about sixteen thousand dollars more for each thirty-second commercial spot even though it ranked much lower in ratings.[14]

As a result of *Murder, She Wrote*'s seeming shopworn quality, programmers at CBS did what many viewers found unthinkable: in the summer of 1995, they announced that Jessica Fletcher would be moving to Thursday nights, where she would compete head-on with NBC's highly successful Generation Xer situation comedy *Friends*. Even if CBS somehow could win by sheer numbers in this contest, it could never capture the valuable young demographic group that advertisers so desperately seek. In *Murder, She Wrote*'s 8 P.M. place on Sundays, *Cybill*, a baby boomer sit-com, would now follow *60 Minutes*. The impending death of TV's newest murder victim, *Murder, She Wrote* itself, was widely predicted in the press.[15] Murder, the nice drawing room variety, was all but dead.[16]

Even as the program floundered on network television, the USA network recognized its power to hold its faithful audience in syndication. For several years until the fall of 1995, USA had been carrying *Murder, She Wrote* reruns in the mornings and evenings, five days a week. When CBS bumped *Murder, She Wrote* from Sunday nights, the cable channel added a sixth evening telecast in the old CBS time slot. It produced a snappy promotion with intercuts from the program. A narrator intones, "Jessica Fletcher, an American tradition?" In the accompanying shot, Lansbury, as Jessica, thoughtfully answers, "Oh, yes." Then the narrator explains that the program is still available to viewers in the same time slot as always, now on USA. The narrator concludes: "The tradition continues!"[17]

At the same time, less ritualized mystery shows, mostly British imports such as PBS's *Prime Suspect* and *The Singing Detective*, succeeded in collecting audiences—among them younger viewers. Their success came partly because they were not the same old, comfortable slipper that *Murder, She Wrote* was; for one thing, they were limited-run series. Their drama often was coarse, their detectives grittily unhappy. The tale could be complicated. And often they broke with a couple of the strongest conventions of the classical mystery. Their victims were sometimes quite young, sometimes innocent, and did not "deserve" to die. Their murderers sometimes displayed psychopathic or sociopathic tendencies. In *Murder, She Wrote*, the story usually was about greed or lust or revenge or something else plainly comprehensible—and it was between adults.

Lyn Thomas wrote about the allure of one highly regarded mystery series, *Inspector Morse*, for fans in Britain, where the show is produced. Thomas suggested that the series signifies "quality television," especially for young women fans from the middle class, for three reasons. First, it plays on picturesque settings that conjure nostalgia for the kind of Englishness that female audiences have come to romanticize in such popular texts as the mystery novels of Agatha Christie; these rich, pastoral, and, at times, elegant depictions are assured through, among other means, the high production values of the *Morse* series. Second, Thomas contends that the show amplifies the ideology of Britain's Thatcher years as it works with a steady hand to contain a disturbing array of postmodern political and psychic anxieties and to reinstate unfailingly the conservative moral order. Finally, Thomas finds *Inspector Morse* contentious and even ambivalent in its gender representations, suggesting at once that the hero Morse is both an old-fashioned, traditionally moral romantic and a "new" man who has learned to nurture and face his own vulnerability. Thomas's textual and audience analyses demonstrate the resonance of Morse as an example of successful "postfeminist" television for younger women—her interviewees were women, generally in their 20s and 30s, who actively worked to negotiate some feminist voice within the text of a series that is not explicitly feminist. It was critical to Thomas's study that these younger women were able to define a "quality" mystery in such terms as complicated gender representations.[18] Although many older women, including one of the women in Thomas's study and many of the women I have interviewed, like *Inspector Morse*, too, the muddied gender identities that its scripts portray are not chief among the reasons. For older women, the nostalgic Englishness guaranteed by the show's high production values and the assured victory of a conservative ideology are "quality" ingredients in *Inspector Morse*, but these women are more comfortable with subtler challenges to gender representation than *Morse* at times provides. *Murder, She Wrote* provides a similar nostalgic Englishness and conservative ideology. (Significantly, Angela Lansbury is a native Englishwoman, and several of the show's plots involved Jessica's travels among relatives in the British Isles; her "home base" is a seaside New England village where Miss Marple would have felt quite at home.) The program also features a female protagonist who, although assertive and successful, operates foremost as a lady—Morse's grit, self-doubt, and anxiety are absent in Jessica Fletcher. Like many of the women she appeals to, she practices a measure of feminism but would never dream of identifying herself as a feminist. *Inspector Morse* may be the new "quality" mystery; *Murder, She Wrote*, while it projects a "new" image of vitality about older women, is decidedly the old "quality" mystery after all.

As far back as 1976, John Cawelti, writing about literature, offered aesthetic reasons for the decline of the mystery genre: its highly articulated structure has made it resistant to change, which it needs to maintain appeal among audiences who eventually grow tired of repetition.[19] It is no accident that classical mystery and detective programs (such as *Murder, She Wrote*, *Perry Mason*, *Columbo*, *Matlock*, and the like) tend to skew toward older audiences, who may find the genre comfortable compared with more experimental programs. Programs with their roots in the mystery, lawyer, or detective genre, such as *Moonlighting*, *L.A. Law*, and *Twin Peaks*, were able to attract younger viewers to various degrees because they broke from hackneyed formulas of the past. Their creators radically fused forms (closed-ended narrative of the lawyer show plus open-ended soap opera, detective show plus melodrama), experimented with radical textual modes (as in *Moonlighting*'s frequent winks at the audience through direct address), or defied the rational logic that traditionally has bound the mystery/detective/lawyer genres (as in dabbling by *Twin Peaks*'s Agent Cooper in the supernatural for assistance with his solutions). By upsetting traditional hierarchies, such shows have struck other chords with younger audiences, as when *Moonlighting*'s female Maddie Hayes is male David Addison's boss, but he constantly presents juvenile challenges to her authority.

The classical mystery, on the other hand, is inextricably linked to the traditional detective and lawyer genres. The original *Perry Mason* series appearing in the 1950s and 1960s (and the revival series of the early 1970s), as well as the *Perry Mason* television movies produced in the late 1980s and early 1990s, is a lawyer show in the traditional mold. It follows a mystery formula very close to that of *Murder, She Wrote*: the hero encounters a dead person and a wrongly accused bystander, then goes to work uncovering clues about a handful of guest-star suspects, resulting in a climactic revelation of the real killer's identity. More often than not, the killer confesses. Order is restored, and relations are normalized. *Matlock*, a traditional lawyer show that has been popular among older viewers during the *Murder, She Wrote* era, follows a similar recipe.[20]

Murder, She Wrote's overwhelming popularity with elder viewers did not brought it critical acclaim. Except for its emphasis on an aging female character, it is just another example of its genre. Its faithfulness to its type is, for the purposes here, what makes it worth studying. The program's text is encrypted with many of the same qualities that apparently attracted audiences to *Perry Mason* and other shows, and these factors will make for a useful comparison. The people I interviewed in connection with this research found those common qualities appealing, but, one by one, they identified what they found to be the program's key distinguishing feature—the Angela Lansbury character. In what follows I offer a textual perspective to show how *Murder, She Wrote* and its predecessor, *Perry Mason*, are encoded to strike ritual appeal with audiences. Then I turn to *Murder, She Wrote* fans' explanation of the show and its role in their lives.

Perry and Jessica: The More Things Change . . .

Perry Mason could hardly be considered "high-quality" or even realistic television by today's critical standards: its stark mise-en-scène, developed for television in its youth, lacks the richness of latter-day lawyer and mystery shows, for example. Its dialogue reflects a fifties naïveté and utter white-male dominance. The character of private detective Paul Drake, for instance, is positively portrayed as a straight-arrow type; at the same time, he gets smiles from secretary Della Street when he refers to her in such terms as "Doll."[21] *Perry Mason*, in fact, with its pat, unified plotlines and highly formulaic structure, seems corny when weighed against the messy drama of, for example, *Law and Order*, NBC's long-running cop-and-lawyer show of the 1990s.

Max Collins and John Javna observe that *Perry Mason* extends "beyond formula into ritual," and we can perceive *Murder, She Wrote* in similar terms:

> A horrible human being, despised by one and all, is murdered; accused of the crime is an innocent who finds her (or his, but usually her) way to the office of defense attorney Perry Mason. Mason, detective Drake and secretary Della Street set out to solve the crime in private-eye fashion, talking to witnesses, examining the clues, dueling with the police, led by the luckless Lt. Arthur Tragg.[22]

Collins and Javna go on to describe the ensuing courtroom battle between Mason and his nemesis, prosecutor Hamilton Burger, and Mason's unmasking of the real murderer, which usually forces out a courtroom confession. *Murder, She Wrote*'s chief differences are Jessica Fletcher's amateur status and, thus, lack of an investigative staff, her frequent travel to different locations (Mason was based in Los Angeles and generally solved his mysteries there), her unmasking of the murderer in an informal setting, and the unusual age and gender status of her character.[23] *Murder, She Wrote* is the first long-running mystery-detective series to feature a woman in the leading role, and no such series has featured an older woman as the central character. Older men (*Matlock*, *Columbo*) are more common.

As a result of this irregularity, Lansbury's role—and its obvious point of contact with the elderly female demographic group—received generous play in popular media. For example, mainstream magazines such as *People*, *McCall's*, and *TV Guide* all gave generous coverage to Lansbury's weight loss, face-lift, and fitness video in the late 1980s. These magazines all took up the slant that Lansbury was "successful" in "fighting" old age while earning the dignity and grace that went with the late years of life. It seemed obvious to disseminators of such publicity and journalism that older women had felt starved for their own kind of heroine—and hungered for information that revealed the model she represented for them. The concentration on Lansbury's cosmetic surgery and weight reminded these women that, in the United States, it is all right for a woman to be old only if she looks young. By contrast, in *Perry Mason*'s heyday, the media paid little attention to Raymond Burr's battle against overweight. When Burr came back as an older man and did the *Perry Mason* movies in the 1980s and 1990s, he was presented as an aged person with a naturally disobedient body—white haired, overweight, in need of a walking cane. This contrasted sharply with network orders to Lansbury: Lose weight—and the lines in your face.

While Jessica Fletcher's distinct characteristics of age and gender are important to viewers, other aspects of the series' deep structure that make the program similar to *Perry Mason* are integral to its success with older female viewers. In both, for instance, little series development takes place, and the main characters do not change much because of their involvement in the "stories" of the series. For example, Perry Mason does, over the course of the series, accrue status as a reputed legendary trial attorney, and he also assumes growing familiarity with the supporting characters as the series goes on. However, the nature of the character relationships does not radically change. Neither Perry nor Della, for instance, is shown to have a life outside the office, and an implicit hint of sexual tension between the characters is suggested throughout the life of the series. The familiarity between the two deepens, but they never consummate their affection. Although there may be a few exceptions to the rule, this deferment of intimacy was born out of two conventions—the producers' strategy to achieve and maintain sexual tension as a story line device and the celibate nature of dramatic characters on television in the age of the repressive 1950s and early 1960s.

Likewise, the *Murder, She Wrote* series does not have much of a memory. A few functional changes occur, including minor cast changes. (For instance, Cabot Cove got a new sheriff when actor Tom Bosley got his own mystery drama and the character of Sheriff Amos Tupper was eliminated.) A few theme-related changes contribute to the growth of the Fletcher character. For example, Jessica changes, gradually, from a dowdy, small-town mystery writer clinging to her manual typewriter and bicycle to a sophisticated, computer-operating New Yorker who gets around mostly in cabs and on jet planes. Her character, that of a well-mannered, pleasant, and curious woman with excellent judgment, remains intact, however. Like *Perry Mason*, *Murder, She Wrote* limits the realm of the personal to the extent that Jessica Fletcher almost never engages in a social relationship for story purposes other than advancing the mystery plotline.[24] She has expressed a romantic interest in a man only a handful of times. Of these men, one turned out to be the murderer and had to be sent, through Jessica's efforts, to prison. In an unusual development, the character turned up on an episode during a subsequent season only to become the murder victim.

It has been demonstrated by many scholars that women and men often prefer different genres because the programs within them are designed to appeal to audiences on the basis of their gender. This line of scholarly inquiry mostly has been concerned with women and men as young people. For example, we know that soap operas tend to appeal to women because they offer extremely open narratives, messy and proliferating story lines, and a high degree of story redundancy that meshes with the repetitive homemaking functions that women often perform.[25] On the other hand, classical cop shows, with their chase scenes and individualist heroes that are framed within extremely closed narratives, tend to appeal more to men, whose role it is to compete and to complete work outside the home. It is not so easy to bracket the genre preferences of older women in pure terms of gender. Their days are not filled with the same repetitive homemaking tasks and constantly renewed nurturing functions that might have characterized them in their younger years. As women get older, they, like older men, seem more occupied with resolution—with completing life's projects rather than with starting everything over again each day. Of course, to a certain extent, everyone's life involves forever sequencing tasks: doing the dishes, paying the power bill. But, for older women, the sheer magnitude of life's repetitions has been noticeably reduced as the women themselves inch closer to their own final resolution. Their repetitive functions turn mostly toward taking care of themselves—the pills, the budgeting of the Social Security check. They no longer cater to what seemed like the rest of the world. These women are not making the same peanut butter and jelly sandwiches, stuffing them into little brown bags. They are not facing the enormity of housework or social performances. They are no longer, probably, questioning how their romances will turn out. They are closing off life's corridors, shedding possessions, paying off notes, finally taking the cruise. So, while they continue to need ritual in their lives, the ritual of soap opera does not necessarily resonate as strongly for them. They might instead crave rituals of assurance and resolution, the latter of which traditionally has been linked with a male mind-set.

For this reason, programs like *Murder, She Wrote* and *Perry Mason* appeal to older women viewers. They feature the process of working out a problem that is comfortingly manageable. This classical mystery genre blends gendered styles—a traditionally male emphasis on closure and a traditionally female emphasis on process.[26] Using the examples of *Cagney and Lacey* and *Hill Street Blues,* John Fiske suggests that some television programs can exist as mixtures of masculine and feminine genres. He cites, for example, these programs' tendency to equate in significance the process of the story with the episodic outcome.

The examples of the mystery genre explored here can also be submitted as blends of the masculine and feminine forms. Neither of the programs emphasizes the physical, with the general exception of the murder event; both emphasize the mental process of solving the crime. Yet, in both series, the climactic moment occurs when the hero exercises cerebral superiority over the offender and, effectively, captures him or her in a snare of logic. Perry or Jessica, in verbally upbraiding the villain and freeing the distressed damsel (female or male) who stands hopelessly accused of the crime, is no less the virile hero than if he or she had chased the villain down in a red Ferrari. Still, the process is more mannerly. Perry does his battle from the defense table and Jessica, often, from the dinner table. Both become involved in feminine-styled discussions about motives, opportunities, and clues, whereas heroes from more masculine-oriented shows, such as *Mannix*, frequently found themselves involved in physical conflict—perhaps being confronted with having to kill the murderer out of self-defense in the climactic scene. Protected by the courtroom, Perry Mason rarely finds himself with his life threatened by the murderer. Occasionally, Jessica Fletcher confronts the villain privately and has her safety threatened, but she is almost never involved in a physical struggle.

On other levels, however, both *Perry* and *Murder* reflect the traditional terrain of masculine texts. For example, a feminine-style text about murder might be expected to dwell on the emotional consequences of the crime—as in a made-for-television movie. Perry Mason never displays emotions about the victim's loss, going straight, instead, into the crime-solving process. Jessica Fletcher, upon discovering "the body," often briefly grimaces, then, like Mason, sets about solving the puzzle. Many times, she has appeared in a black dress at a wake or funeral but only to advance the plot by collecting clues or confronting a suspect there.

Also reflective of masculine genres are both series' tendencies to employ agents of white patriarchy, unmistakably good guys, to orchestrate the restoration of order from chaos. In the heat of investigation, however, both heroes explore the boundaries of legitimacy in order to find "truth." This type of action, concerned more with rightness than with authority, may suggest a feminizing of the text. Perry Mason often makes decisions that temporarily cause the authorities around him to question his ethics. He skirts the law on such matters as tampering with or withholding evidence, failure to report a homicide, and perjury—offenses that, on their face, clearly oppose cultural values. Truth and justice prevail, as part of Mason's design, but the story encourages us to consider opposing values—an innocent person's right to freedom versus adherence to the letter of the law.

Perry never breaks the rules but bends them to suit his needs in a ritualistic game of wits against the authorities who have the wrong killer. In one episode, for example, a quintessentially helpless maiden approaches him for assistance, which she needs urgently, but she is financially embarrassed. Perry asks her how much change she has in her purse. "Thirty-eight cents," she answers. He takes it and declares himself legally retained as her counsel. (An oppositional reading of this exchange might see some irony in a lawyer taking a poor "victim's" last thirty-eight cents.) Occasionally, Perry, having unmasked the true killer, shows compassion for the person, with whose plight he sympathizes. His initial case won, he now will offer his services as defense attorney to the real killer. (Of course, the viewer never sees *those* cases.) Jessica Fletcher, in *Murder, She Wrote,* has on numerous occasions exercised similar compassion for the offender whom her sleuthing has identified. For example, in one program from the late 1980s, the "murderer" turns out to be a young, pregnant Amish woman who killed her evil lover with a pitchfork in self-defense. In a common ending, the mystery concludes with Jessica, having uncovered the truth, happily assuring the woman that the law would be lenient with her. (Of course, Jessica does not stick around to help the woman navigate her course through the legal system.)

Another way in which *Perry Mason* and *Murder, She Wrote* represent a blend of masculine and feminine genres is their heroes' blend of deductive reasoning and intuition to solve crimes. Both rely more heavily on logic, the traditional realm of male thinking as made part of the popular lore through the Sherlock Holmes stories of Arthur Conan Doyle. In a typical instance, Perry realizes the identity of a killer because Lieutenant Tragg mistakenly suggests that Mason has accidentally tracked a feather from the murder scene into his office. Immediately, Perry knows that the female visitor who left earlier must have brought the feather in on the bottom of *her* shoe and therefore must be the murderer. In a *Murder, She Wrote* plot that, coincidentally, also involves a feather, Jessica figures out the killer's identity when she realizes the reason for a feather found (by her, of course) on the corpse. On the other hand, both Perry and Jessica intuit as well, although Perry relies on secretary Della to do much of his intuiting. These three characters—Perry, Della, and Jessica—all seem to be able to tell when someone is lying to them and, perhaps more important to the story, when the falsely accused person is truthfully professing innocence. What is more, no other characters on these two programs, especially law officers, seem to have this ability.

Murder, She Wrote achieves this blend of feminine and masculine textual tradition on another level. The show is highly episodic, but it does offer occasional rewards for regular viewers. Recurring characters, such as nephew Grady, do appear, and occasional reference is made to his earlier "problem." Michael, a mysterious British MI-6 agent, occasionally appears, and Jessica always chides him for getting her into a previous jam. Such references award longtime, loyal viewers with a special intimacy and tend to focus, for a moment, on Jessica's life as process. Still, these references are so simple and brief as not to confuse the naive or forgetful viewer who may be concerned only with the current story line.

Murder, She Wrote bears a feminine style in another way as well. Jessica Fletcher is not an officer of the court like Perry Mason; she is an amateur.[27] She is an older woman who bravely speaks up when "the system" is about to punish the wrong person for a grave crime, and she successfully challenges authority. In doing so, however, she is almost self-deprecating, telling various lawmen, "Of course, I wouldn't dream of telling you how to do your job, but . . ." or "I just write mystery stories, and I'd like to help." She ably leaves the murderer in the hands of the authorities, reinstating their legitimacy after metaphorically winking at us behind their backs. Then, she returns to her home in Cabot Cove (or New York) to whip out another harmless mystery novel.

Jessica Fletcher, although contemporary, bears a striking resemblance to the heroine of so many Agatha Christie novels, the properly English villager Jane Marple. Miss Marple is single, a gossipy busybody, yet preeminently gracious and civilized. Early comments on *Murder, She Wrote* in the popular press directly compared the two. Lansbury herself described Jessica Fletcher as "an American Miss Marple."[28] Lansbury's having played Miss Marple in the film *The Mirror Crack'd*, released in 1980, facilitated this comparison. Differences between the two characters are also striking, however. Significantly, Jessica had been widowed by the love of her life; she is, as biographers Rob Edelman and Audrey Kupferberg have noted, "nobody's stereotypical maiden aunt."[29] And, although she is cultured, like Miss Marple, she is neither eccentric nor dowdy like her. Jessica's Cabot Cove community seems to depend on her not just for solving murders but for administering civic business and expediting community fellowship. Plainly said, she is an activist.

But Jessica Fletcher and Miss Marple have enough in common that it was likely that if an American reader had been a Christie fan, chances are she would enjoy *Murder, She Wrote*. Among the elder *Murder, She Wrote* fans I interviewed, most reported that they enjoyed reading Christie novels if their eyes could tolerate such activity. Many had read the novels in middle age, and most preferred them to the works of other famous mystery authors who were males or to contemporary mysteries by women. "A good, clean drawing room murder—that's what I like," one woman explained to me.

A Good, Clean Murder and Its Viewers

In 1995 and 1996, I talked with twenty-five women from Wisconsin about their longtime interest in *Murder, She Wrote.* These women had disparate backgrounds, ranging from working class to upper middle class, with urban, suburban, and rural lifestyles, and including African American and white Jewish and Christian ethnic identities.[30] I met most of these people at senior centers and assisted-living homes, where workers helped me identify self-described fans of the show who were willing to talk.[31] (Included here also are brief summaries of the *Murder, She Wrote*–related views of other elders whom I interviewed about their general television viewing interests. These included one multiracial gay man, two white lesbians, and one straight American Indian woman.) The people I met overwhelmingly authenticated the notion of *Murder, She Wrote* as a ritualized text.

What surprised me was the range of perspectives about the show and how passionately the different women felt about it. For example, all the women cheered the program's representation of a strong, intelligent older woman performing important actions in public ways, but there were splits along lines of social class and ethnicity about how to interpret the character's locus of strength. What united all these fans was the view that *Murder, She Wrote* broadcast good publicity for what they saw as unchanging ideals. In doing so, the show was a salve to relieve their anxieties about the decaying morality of entertainment television.

Also apparent in the women's stories was a tendency to channel their personal strength toward allowing themselves to routinely incorporate *Murder, She Wrote* into their lives. These were women who, in one way or another, had been accustomed to submerging their own pleasures in favor of awarding others theirs, so it was heartening to see that watching *Murder, She Wrote* provided them with such an intense and faithfully anticipated reward that they were willing to restructure their sense of the ordinary to ensure the experience. In that sense, it seemed that the protagonist's assertiveness fanned the flames of their own. For these elders, the program was something they almost unfailingly counted on.

The secular rituals that can be a part of everyday life can be easiest to perceive when they are disrupted. The people I spoke with told of their anger when CBS repositioned *Murder, She Wrote* from Sundays to Thursdays. For example, Paulette, a woman who had lived on a farm all of her life,[32] brought up the topic soon after her group meeting about the program was convened: "I'm so mad they moved her. Every Sunday night for years, I have gone through the routine of feeling that *60 Minutes* is over and now here comes Jessica! It just made my week."

Many similar quotations appear in my interview transcripts, which are taken mostly from meetings with various homogeneous groups. The women said they had looked forward to their favorite program every Sunday evening, that it was a pleasant way to mark time, and that the program represented a set of ideas that they enjoyed seeing celebrated over and over again.[33] They amplified their mutual conviction that they were not looking for surprises in form and content, only for surprises in the variations within the predictable plot structures. Mystery dramas seemed to represent a particularly homologous fit for older people, who enjoy a comfortable routine and confirmation of long-held beliefs. I want to be clear, however, that I am not suggesting that older people have little sense of adventure: I have noticed in my discussions with elders an interest in trying new concepts and learning new information, but it seems to me that they tend to insist on these pursuits within very familiar boundaries. New technologies, for example, diffuse to younger populations well ahead of older ones; elders may well adopt the technologies but want to learn something about them and may take a while to consider their decision. They want to feel comfortable. They like to feel the land underneath their feet, so to speak.

Murder, She Wrote offered this kind of stability spiced with change. Not only did it introduce diverse and exotic settings, when Jessica would whisk off for a quick stint in, say, Moscow, County Cork, Honolulu, or even the barrio of New York, but it introduced diversity in a nonthreatening way for its predominantly white audience. Members of ethnic minorities, who were portrayed sparingly, were almost always presented as clean-cut (read white) all-Americans. During the life of the entire series, only a few African American men had major guest roles on the show; these characters might well have been played by Bryant Gumbel or Harry Belafonte. The black jazz performer from New Orleans might have killed his mistress in a crime of passion, but she would never have succumbed to street violence in the manner that *NYPD*

Blue's homicide victim might. Minorities were framed as "safe"—middle class—free of the class-based tension that so often permeates narratives about minorities in messy police crime dramas. *Murder, She Wrote*'s diversity was diluted. The one musical character who appeared in dreadlocks, in one of the show's latter episodes, "The Sound of Murder," crucially was a folk singer (à la Tracy Chapman), not a gangsta rapper. This watered-down diversity was not noted by any of the elders I met, but the members of minority groups tended to say that they liked the fact that the show portrayed these groups and wished that it would have done so more often. In the case of African Americans I interviewed, I had to wonder whether my being a white researcher might have inhibited them from more freely criticizing *Murder She Wrote*'s scant portrayal of ethnic minorities.

Likewise, the show, especially in the Manhattan episodes during the latter half of the series, presented new technologies and newfangled concepts to its audience in an adventurous, "user-friendly" manner, as if they were stops on a movie studio trolley tour. For example, plot structures involved facile explanations of answering machines, cellular phones, personal computers, rock music video production, virtual-reality game construction, and various other "mysterious" aspects of the communication industry that more typically are associated with youth than with older people. Several of the women I interviewed said they liked these kinds of introductions, some of which helped them adopt new technologies on their own and others of which merely helped them understand as much as they wanted to know about technologies that had no personal application for them. Jessica went from stubborn allegiance to her manual Underwood to happy partnership with her personal computer over the life of the series, stopping to relate the workings of newer technologies to her audience through the plots along the way. For example, Jessica caught one murderer (portrayed by Ralph Waite, who had played John Walton, the father in the historical melodrama *The Waltons*) by tricking him with the murder victim's redial button, a solution later similarly employed by Ben Matlock. In the climactic scene, Jessica told the murderer that perhaps he had not known what a redial button was but she did, and she proceeded to explain cursorily that it was a button on the telephone that repeats the dialing of the last number called. Whenever Jessica found herself in high-tech environs, she was portrayed as confident and comfortable, without exhibiting stress at learning the technology. For instance, the premise of one show centered on one of Jessica's novels being adapted into a virtual-reality game. Scenes showed a self-assured Jessica donning the special eyeglasses for using the technology and deftly finding her way around the game. Before long (after all, the murder had to take place and be solved in under an hour), she was freely issuing commands to move certain features around in the game or omit them entirely; she had been able to master the technology to the point of assuming the roles of critic, user, and designer. And, of course, her ready understanding of virtual reality clinched the solution of the "real" crime on the show.

Murder, She Wrote curried ritual involvement with its audience quite intentionally. Such moments as its instructive mode of address about technologies were but one of its points of interpellation for its audience.[34] The program, it seemed, knew with whom it was dealing: people who were not at the forefront of popular trends, who were, in fact, conservative in many ways, but who were curious on perhaps a small level about how things worked in the world (at least the world of the middle class). For example, Naomi, an African American resident of Milwaukee's north side, told me:

> I got inspiration from her, because, even though she was an older person, she could change from one way to the other. She could go from the old technology to the new—for old people that's a possibility—you can change. And not only with technology. It's the way she dressed and the way she carried herself, you know. She was classy. I always paid particular attention to her hair, her combination of clothes. And yes, 'cause that's what I like to do, too. Like learn the computer in my 60s. She learned how to do the modern-day kind of thing.

The viewers' diverse backgrounds informed their views differently on many minor fronts, but on the large issues these fans displayed an amazing unanimity of interpretation. Clearly, this was a program whose makers understood their audience, and they created a text that their audience understood just as well. It is useful to picture the *Murder, She Wrote* fans as an interpretive community in the same sense that we can picture other media audiences as interpretive communities, following the suggestion of Tom Lindlof and the empirical work by such people as Janice Radway and Henry Jenkins.[35] Here was an audience whose members bonded wholly with the program, who found the program to resonate thoroughly with their worldview.

As an interpretive community, it was not necessary that they knew one another personally, but they shared, in large part, a core belief system that linked them with the messages that Jessica Fletcher and her fictional life represented for them.

Murder, She Wrote Fans

My goal was to get to know the views of different kinds of women. I could not possibly meet women from *every* social background, but, by meeting women from several different social groups, I would be able to say something about the general resonance of *Murder, She Wrote*'s themes as well as the different ways in which viewers inflected these themes with their own experiences.[36] The first group I met consisted of five residents of Lakeside Christian Home. We met twice in the sunroom of their sprawling retirement home overlooking Lake Michigan in a tony section of Milwaukee's old east side. Living expenses at this assisted-living home were in excess of two thousand dollars per month and included the cost of basic cable television. (No one in the group subscribed to premium channels.) The women had lived for varying lengths of time at the all-white center, where the ambience tended toward the formal, with men appearing in sport coats in the daytime and the women in dresses with nylons. In the formal dining hall, where servers took orders for lunch and dinner from a limited menu, grand woodwork and original art and antique pieces matched the European architecture and furnishings of the rest of the building. The women in the group all had college degrees and had all been married to college graduates; several had had parents with college degrees. Several had had their own careers, and a couple had been homemakers. Some grew up in the area; some had moved from other urban or suburban locations.

I conducted the interviews with the assistance of two graduate students, and we were all received warmly, with many questions from the group about my work, the students' plans, and their Taiwan homeland. An employee at the home set up the first meeting for me, but I discovered, when setting up the second one myself, that these women were not easily assembled. They, of course, had a few of the stereotypical elder activities that kept us from meeting at times—doctor visits and funerals, for example—but they also had canoe trips and continuing education courses.

Five African American women joined the *Murder, She Wrote* interview group at Northside Senior Center in a rundown section of the city that bears the markings of violence and poverty. The center and its furnishings are old but not grand—lots of vinyl and Naugahyde. Off to one side of the large main room, which serves as a congregate meals center, is a medium-size television set, with the volume perpetually turned up loud. Here, residents might watch their favorite quiz show or news program or, as a few ritually did, a daily syndicated episode of *Matlock*. During the time of our meetings, the program of choice was *The People vs. O.J. Simpson* on CNN; unlike the elders in the group themselves, the center subscribed to cable television.[37] The group met in a small room off the main hall where holiday decorations and craft project materials were stored. The group members did not know each other despite their frequent visits to the center. They met three times, and I visited one group member, Naomi, in her north-side home for a one-and-one-half-hour follow-up interview. The group members, except for one, had worked as unskilled laborers. All were widowed or divorced.

Because a few of the group members were in poor health, it was difficult to get them all together for a meeting. A couple had to cancel the day of scheduled meetings in order to go to the doctor. One of the women dozed off during a meeting; she was one of the eldest members of the group, and her health was failing. A graduate student conducted the group meetings and brought homemade pie and other refreshments. This tickled the group members, who said they appreciated so much attention being paid to them. Rarely, as one woman pointed out, do younger people ask elders about their views and listen for long periods. This group seemed to be one of the least bound to accept a fellow speaker's viewpoint—ample disagreement helped frame the discussions.

The group of women in rural northern Wisconsin met twice in a conference room of a public library in small-town Fond du Lac, near Green Bay. None of the group members knew one another before the meeting, but they all were acquaintances of the parents of the graduate student who recruited them. Several came from their farm homes far outside the city, and a few lived in retirement apartments nearby. All were white. They had varying education levels, and all had worked as farmers (an occupation they tended to list as "farmer's wife"). Most had worked in factories to supplement the family income at one time or another.

These women tended to have the freest schedules. Most were in good health and had few social commitments. Among the women who still worked on their farms, their days were somewhat flexible, and they willingly accommodated the interviews.

The Jewish Senior Center is part of a large complex serving the metropolitan Jewish community in Milwaukee. Elders were bused in to the center, on Milwaukee's east side, daily for classes, the midday meal, and other services. Participants ranged from lower middle to upper middle class. Most were urbanites. The group met in a large cloak room, with banquet chairs positioned in a circle. We understandably felt a bit jostled as people stepped in to hang up or retrieve coats and unfailingly asked what the group was doing. The group met twice.

Among frequenters of middle-class suburban senior centers, I formed a small group that met twice on our university campus. The three participants were all college-educated and were homemakers with husbands who had worked as professionals or in business. It was extremely difficult to recruit members from these locations, although there was no shortage of self-described *Murder, She Wrote* fans. Most people said they were too busy with hobbies or volunteer work, while others said they felt hesitant to share their views without their husbands present. A few said they were embarrassed to admit they were television fans, and I suspected this was the case among others, who did not want to link themselves with an activity that they understood to be looked upon as trivial by many of their contemporaries.

The women who wound up participating were very busy, with a couple of them still working, at least part-time. They looked upon the group as a "fun" opportunity to talk about a program they enjoyed, although, on the whole, they were the most critical group with regard to television generally, saying they found much of its content offensive. These women watched only a few hours of television weekly, according to their statements.

Jesica Fletcher-Angela Lansbury

Invariably, when asked what drew them to the program, the *Murder, She Wrote* fans quickly volunteered either or both of two answers: the star/character and the show's lack of sex, violence, and rough language. Jessica Fletcher herself, or Angela Lansbury, was the most popular response. What differed among the fans mostly is whether the character or the actress stood out more, and this difference often played out along class lines. When women were asked what first suggested to them that they might enjoy the show, they most often said it was the star. "I've seen her in other movies," said Yvette, a farmer with a high school education. "She's very talented. She has lovely legs." Women also said they had been drawn to the show because they enjoyed mysteries. Among other favorite mysteries had been *Matlock, Perry Mason, Columbo, Cannon,* the *Agatha Christie Mysteries, Ellery Queen,* and *MacMillan and Wife.* These fans resoundingly reported liking *Murder, She Wrote* best of all. They all preferred Lansbury's "ladylike," "up-to-date" Jessica Fletcher to Andy Griffith's folksy Ben Matlock, Peter Falk's crusty Columbo, and Joan Hickson's "Oh-dear!" Miss Marple.[38]

Angela Lansbury, the Actress

Lansbury herself was the most attractive draw the show had to offer most of the fans from higher-status backgrounds. The upper-middle-class women volunteered memories of Lansbury's stage and film career over the forty years prior to her television series' origin. These women, within fifteen years of Lansbury's own age, recalled seeing the star's films and, in a few cases, her Broadway and London stage appearances.

Beyond this, these more affluent women seemed knowledgeable about Lansbury's handling of her career. "I wasn't surprised at all [when Lansbury got a television show], because she is intelligent, ambitious, and always wanted to be a director," said Bea, a Protestant home resident. Several of the women in these groups mentioned that Lansbury's brother, Bruce Lansbury, and several other relatives worked on the show. Some had read press accounts and knew the relationships between the actress and her family member employees. A couple of fans noted approvingly that Lansbury had interrupted her career several years before the television show so she could move her family to Ireland in order for her teenage children to get away from their involvement with the U.S. drug culture.

Most fans spoke of the actress's appearance as worth aspiring to, but one of the women in the group of upper-middle-class Protestant elders, Helen, remarked critically: "I didn't approve when she had her face-lift and her makeover [during the show's early years]. I like people to be as they are.... Some of us have had a problem gaining weight, and I like to see an older person on television who has gained a little weight, who doesn't have a youthful figure."

Although the upper-middle-class women spoke of Angela and Jessica as separate from each other, they said they "lost themselves" in the character. Sharon engaged this slippage when she related how the program spoke to her about the role of older women in society: "I was a housewife much of my life and then went back to work when I was older, like Angela seemed to do after her husband died." Sharon was aware that Lansbury, the actress, did not go back to work when widowed but that Jessica Fletcher, the character, had done so. Even so, she creatively blurred the identities of the actor and the character, just as she creatively engaged in self-identification through them.

These more affluent and well-educated viewers tended to mention one of the middle seasons of *Murder, She Wrote,* when Lansbury took a reduced role, appearing only to introduce and close the stories that featured other detectives during many of the episodes. This near Jessica-less season left the fans feeling frustrated with the substitute programs that they said mostly featured men. "It didn't appeal to me," said Donna. "I was waiting for her to come on, to do something, and she never did it," added Helen.

It seemed important to the middle-class viewers that Lansbury seemed "down to earth." They found this quality easily transferable to the character Jessica, who, despite the trappings of her success, came across to them as "just plain folks."

This slippage between actress and character sometimes extended to the more middle-class viewers as well. For example, in my meeting with the members of the Jewish Senior Center group, some of the women talked about their admiration of Lansbury and her character even though they could not personally identify with her:

Lillian: I was born in a different time. When I was young, the boys were sent to get an education. The girls were sent to work. . . . You never see *her* in a kitchen!

Rose: I don't think Jessica has time to cook. She's too busy.

Nina: I think they even make the pot of coffee before she gets on the set!

Such instances of confusion, as this one involving the blurring of Lansbury and Fletcher, were common in my conversations. Some centered on misunderstandings about the show's scheduling and whether production had ended. Once in a while, someone in the group would catch the slip and more or less impatiently get the group back on track. When a group member just didn't get it and remained confused, sometimes other group members got angry. This was especially true among the oldest women, who may have begun to feel mounting trepidation about being categorized as senile. I had the feeling that their anger might have been motivated at least partly out of concern for how one's social group was presenting itself for public consumption through an academic. More often, the slips went unnoticed by group members, as in this instance, when Edith, of the Jewish Senior Center, said she wished that "they" (meaning the network) would allow Angela Lansbury to show her age as her program progressed: "They should let her age as the years are going. To show time elapse. A lot of programs do that, I notice. Perry Mason was one, and there are other ones. I loved to watch Perry Mason. They never did anything, and then one day you saw him in a wheelchair. In an accident. They showed time elapsing. That's what they should be doing with Angela." Edith had confused Raymond Burr's two signature television dramas, *Perry Mason* and the cop show that followed it, *Ironside*. No one in the group seemed to notice, and some group members were nodding in agreement with her.

Although even middle-class viewers occasionally conflated the two, it was much more common for the less affluent viewers not to distinguish between the actress and the part at all. They identified literally with the character. On the whole, these women did not relate their viewing intertextually to newspapers and periodicals as the middle-class women did. The women from working-class backgrounds may have seen some press accounts but did not consider them important enough to remember. For them, the salient thing was the text—the character. They liked absorbing themselves in the character's acts. Gossiping about the actress had little place in their lives. Among the women at Northside Senior Center, Justine commented: "All I know is 'Murder, She Wrote.' It tells you the names, but that ain't what I'm interested in. I'm interested in *her*." Most of the women at the center spoke of the character only as "she," "her," or "the woman who solves the mysteries." One woman called the character herself "Murder, She Wrote" ("I know that Murder, She Wrote is a good person").

One reason for such disinterest in Lansbury herself may have been the working-class women's readier acceptance of their own aging, measured against the pressure that the middle-class women felt to "stay young." None of the less affluent women colored their hair, for example, but some of the middle-class women did. Both groups of women liked Lansbury's "attractiveness," but the working-class women did not tend to brood about their own lost youth. They more likely referred to themselves as "seniors" or some other such tag than did the more affluent women, especially the younger ones, who outspokenly fought such labeling. The working-class women seemed less troubled about the end of life—or, if they were very elderly, perhaps just foresaw less of a future.

An exception among more working-class viewers was Naomi, who was educated with an associate's degree. She did not generally refer to Lansbury by name but was aware of her status as an actress. When I told her, for example, that the part of Jessica Fletcher had initially been offered to Jean Stapleton, Naomi commented that she knew Stapleton to be accomplished in comedy and in other forms of acting, but she felt she lacked the "poise" of Angela Lansbury, and that quality was crucial to the role of Jessica. I asked Naomi whether she thought she might have been drawn more to the series if it had starred a black actress, such as Della Reese. "Not Della Reese," she said. "She would be wrong. But the black actress who played the lawyer one time . . ." "Cicely Tyson," I said. "Yes, Cicely Tyson probably would have drawn me even more to the show. But it was not important to me that the actress was white. I saw her as an older woman, and I was able to identify."

Some of the middle- and upper-middle-class women came to their interviews with press clippings. One of the rural northern Wisconsin women had done some reading in the library to prepare for their group interviews. All these women seemed to be pleased to assist with the research process by quoting authoritative sources. Many said they always noticed when their newspaper had a story about their favorite program. As Brenda, a Jewish Senior Center member, said, "Her hair is dyed, I know it. The papers have said she does it herself. She does a good job, and I think she acts well. And she does know all the answers." Her fellow group member, Lillian, countered:

> I must tell you, I read a thing written by her, and this is exactly what she said. When she started that program, she was not a youngster. They gave her the program, and then they told her they wanted her to have a face-lift, because she was showing wrinkles. They wanted her to keep her hair

dyed, and she was heavy. She was a heavier woman, and she went on a severe diet, and she had a tummy tuck.

The women in the working-class Northside group said they thought the show's star was good at breaking stereotypes about older women. "I think she's better looking than when she first started," Hattie said. "She looks better, dresses better, and is better looking," Mary said in agreement. None of these women mentioned any knowledge of Lansbury's plastic surgeries or fitness program—they spoke again only of the character.

Jessica Fletcher, the Character

The upper-middle-class women said Jessica Fletcher's qualities and work habits were among the main reasons they enjoyed the show. Overall, they especially liked Jessica's being an older woman with a successful career. Donna, at the Christian home, admired "the way that being a writer opens so many doors for her," and Helen spoke about liking to see "a woman who is old and who does a good job." Echoing her fellow interviewees, Helen later noted that Jessica is an "efficient amateur"; these fans liked how the character blended logic and intuition in her work. None of these women projected Jessica's writing career literally onto their own lives but translated it in the abstract: many fans said they drew on Jessica's independence for their own autonomy in making daily decisions for themselves. As one woman said, "Jessica reminds me that I don't have to let my children decide things for me."

The women generally said Jessica inspired their approach to dress. This self-identification was strong for Jane, an upper-middle-class suburbanite who more or less consulted Jessica on how to present herself for different occasions: "I kind of like to see her dress for different groups: Is that what I would wear if I were going to an opening, a theater opening in New York?"

Another suburbanite, Hortense, said Jessica's activities allowed her to travel vicariously to places where she could never have gone in her lifetime or could no longer go because of personal limitations. Similarly, the rural women enjoyed seeing incorporated into plots Jessica's picturesque Maine hometown with its antiques and "quaint" shops, luxurious hotels, the exciting signifiers of Manhattan, and the variety of natural scenery in location shots.

In the series of interviews I conducted with older gay people living in

Milwaukee, the focus was not on *Murder, She Wrote* but on their general views about television, the subject of chapter 5. *Murder, She Wrote* had been a favorite for three of them, one man and two women. These elder gays tended to say they liked the program because of Angela Lansbury's involvement in it but more because of the qualities of the character she portrayed: self-confidence and tolerance. "It's a tough world out there, especially when you're gay," Jean, a lesbian, explained to me. That the show never dealt with gay issues directly did not bother these fans, but a couple of them said they would have enjoyed the show even more if gay characters had been included.

Even as most of the women said they found strength through Jessica, some said husbands were threatened by her. Rose, a fan I met at the Jewish Senior Center, related a story similar to that of many other older married women I interviewed: "My husband didn't like what I watched, so he went and got himself a little black-and-white TV for the bedroom." Often, Rose and her contemporaries reported, their husbands disliked the comedies, musicals, and especially *Murder, She Wrote* that all these women liked. The married women maintained control over the big living room television set while the men took off to watch a smaller set and generally chose sports programming. Together, many of these couples watched evening news broadcasts and game shows, such as *Wheel of Fortune*, before splitting up for prime-time viewing.

Many women said their husbands had enjoyed watching other mysteries with them, such as *Columbo* and *Matlock*, although some husbands avoided the genre, preferring more action-oriented dramas and sports. Several of the women said their husbands did not care for the strong female lead. Lillian, a member of the Jewish Senior Center group, related this anecdote:

> What I like about the program is that I like to see an older woman who is smart, intelligent, and achieves something in this world and writes books and uses her head and her mind and speaks up. However, a funny thing happened on the bus this morning. I asked everybody's opinion on the program, and one woman raised her hand and she says that she has a very hard time with that program. I asked why, and she said, "I have a husband at home and I love the program, and he absolutely detests her. He screams and makes noise all through the program, and he makes fun of the program until I can't enjoy it." And I said, "What do you mean, he doesn't like

the program?" She said, "He objects to any woman who's so smart-alecky that she knows all the answers, and he isn't going to be going for that!"

For these women and others, I perceived an undercurrent of anger toward husbands who might, in the broad sense, express disapproval about their wives' program choices. Some went as far as to tell me that they resented their husbands, upon retirement, attempting to "take over" the purview of the home, which had largely been their own to control during the husbands' work years. Most of these women reported the control over the large console television set as a moral victory, although some did so in benign terms. Several described the battle for the big set as a constant turf issue, except when it came to *Murder, She Wrote.* The program had sacred status, and almost all the women I met told me that their husbands ritually withdrew from the living room after *60 Minutes* went off each Sunday night or they ritually fell asleep a few minutes into each *Murder, She Wrote* episode. "That's *my* show. It's *my* time. Mister don't bother me," said Frankie, an American Indian woman living on Milwaukee's south side.

The rural fans spoke admiringly of Jessica's accomplishments; they were excited by her stature as a novelist and liked seeing her "make the most out of life despite being on her own." Several took comfort in Jessica's new success as she encountered widowhood, as they were doing. These women saw an instructional message in Jessica's needing neither a man nor children. For many women in the cohorts that we currently define as "elderly," the social significance of having a husband cannot be overstated. Most of these women either had not worked outside the home or had jobs that paled in significance—economically and in social status—next to their husbands'. They lived in marriages in which husbands took responsibility for what were termed the major decisions. Husbands most often managed the finances and home and automobile maintenance, and they served as the family bodyguard. Losing one could mean the loss of personal security. It could also mean refreshing liberation. As one widow told me: "I loved my husband. I miss him. But I am on my own for the first time in my life, and there is something to be said for that. I don't want another husband. Jessica does quite well on her own, as you can see. Being a widow is not the end of the world."

The fans showed great interest in Jessica's relationships with people in other social groups. For the Northsiders, Jessica's "work with young people" resonated large. "She sets a good example for older and young people," Mary noted. "I just think there is a need to have somebody like her that will do things in not a violent way but sit down and talk with the young people."

Her parity with and support for young people represented an ideal for the higher-status women, both at the retirement home and in the suburbs. For example, Protestant home resident Bea said, "She respects the young people she is in contact with, and they have great respect for her. . . . That's true of all the people she's in contact with in the show." For Bea and others, Jessica Fletcher proved that elders could both give respect to and earn it from younger generations, and they wished they could see more of that in their own everyday lives.

These fans also said they liked how Jessica "works respectfully with young people" and was able to enjoy many platonic friendships with men. They proudly stated that Jessica "helped" law enforcement when she inevitably corrected the police officer who had chosen the wrong suspect. The suburban fans jokingly said it came as no surprise to them when Jessica unfailingly solved the crime in the wake of the police's haplessness. (The program usually portrayed the police as male, although a handful of officers over the course of the series were women. Cabot Cove authorities were almost all men, including the two police chiefs.) "All my life I've known that women were superior thinkers to men," Jane said with a smile. "Don't let it out of the room," joked Sharon, her fellow group member.

The rural women spoke more critically, as in this comment from Yvette: "The one thing I dislike is the fact that she is above any police or detective. She solves everything. It gives me a feeling that the police are inadequate, which I don't think is true all the time." This group of women, who had spent most or all of their lives on farms, held out the most traditional conservative views of the character. Their own husbands had literally been the boss, and Jessica's constant challenge of patriarchal authority grated against their idea of what was "natural." Ironically, as farm women who sometimes held full-time jobs off the farm as well, they had held greater responsibilities and endured more hardships than most of the other women, but they were the most likely to identify themselves through the men in their lives. They also were the only group to highlight the adequacy of police. This may have stemmed from their own impression of law enforcement's most visible role in their rural environment—protection of property.

This contrasted deeply with a more urbanized view of police—as mitigators of or participants in conflict.

The African American women said they admired the character's competence and superior intelligence next to the police. They attributed her ability to solve crimes the police could not in part to her friendly nature and nurturing relationships with others. These women expressed admiration for the character's problem-solving and peacemaking skills. "Seems to me she solves murders with a kind of ease," Gwen said. "She doesn't have to do a lot of work . . . go looking for solutions that much. They just kind of come naturally. [Yet] she's not in a hurry about it."

By contrast, these women believed the police to be in too big a hurry, rash in their judgment. "The policemen doing a whole lot of running, running, running," Justine said, "and she just uses this" (pointing to her head).

Naomi told me that Jessica's dealings with the police made her think of her own everyday life:

> It shows that the police are not invincible, you know. In the rural areas, or in the suburbs, the police may not seem like reality, but here in Milwaukee on the north side, it's a reality that you look in the face all the time. . . . People see things all the time. And they get used to seeing them. A lot of people don't really understand the things that are a reality for a low-income person.

Naomi recalled a time when she got fed up with having a crack house in her neighborhood and finally called the mayor's office to get some attention turned to it. Like Jessica, she had both worked with authority and challenged it. She liked the way that Jessica Fletcher perceived both good and bad in the police: "I see them not always as a friend or a protector. But I think the police are important." She liked the way that Jessica ultimately related to the police in a positive way, "showing something good for the kids, that it can really happen. . . . She was always collaborating with the police. She didn't try to do everything by herself. And that was important."

Naomi said the Cabot Cove police appealed to her "because, in most cases, police show their authority. But in this particular little place where she lives, they respected her a lot. Maybe it was because she was so down to earth."

The rural women attributed Jessica's superiority to logic and a work ethic, spiced with traditional feminine intuition. They tended to like Jessica's "ladylike" and "gracious" way: "She has a way of really getting across to even the men that are involved with her goings-on and whatever she's involved in most of the time," said Karla. Still, Fran pointed out, assertive is different from aggressive; Jessica is never rude or haughty, and she always respects official authority even if she must "save the policeman from his own ignorance."

Some of the women were passionate about Jessica's assertiveness. For example, Lillian, of the Jewish Senior Center, said:

> You can be older, you can be brainy, you can have a career, you don't have to be deferential to men, think everything they say is the God's honest truth. If there's a man on the program and he says something stupid, she tells him, "This isn't so." She stands right up to them. There have been some real arguments on the program with men who think they can put her down because she's a woman and say she doesn't know anything.

Among the fans I met, there seemed to be a split, not along any particular class or other line, with some of the women saying they wished that "real life" gender relations were as they were portrayed in *Murder, She Wrote* but that they realized the program was something of a fantasy; many others, by contrast, said they found inspiration in Jessica to shape their own relations with men. "You don't have to let a man think he's always right," said Sarah.

Most of the fans volunteered that they especially enjoyed Jessica's understated relations with recurring male characters. Most said they tended to "read between the lines" and see a romance between Jessica and Dr. Seth Hazlitt, played by the earnest, curmudgeonly character actor William Windom. The fans liked to see the friends in scenes involving going out to dinner, dressing up for events, and engaging in light repartee. The more affluent women volunteered Windom's name, the more middle-class women referred to him as either "the doctor" or "Seth," and the least affluent women tended to blur his identity with that of the Cabot Cove police chief, a married character who did not see Jessica socially. Whether these women knew Windom the actor apart from Hazlitt the character did not seem to have substantial bearing on the pleasure they took from the text. The distinction was a mere marker of class difference. A few women in several of the groups said they wished that Jessica and Seth's romance would progress, or that she would have a heavier romance with one of the dashing characters she had met in the out-of-town episodes. Naomi, the Northsider, said she thought of Seth as "a really good, close friend" and believed that the writers would have written a romance into the story if the series had been allowed to go on. Like the other elders, she strenuously objected to the possibility of a sexual relationship, however, stating that this was just the kind of content she avoided by watching *Murder, She Wrote*. Suggestions of a romance for Jessica were not received popularly by fans on the whole. "That would ruin the whole story," said Brenda of the Jewish Senior Center. "This is a wholesome story. It's family entertainment."

When her fellow group member Louise countered: "Women fall in love," Lillian cut in: "Of course, they do. But we don't *need* her to fall in love." For fans such as Lillian, Jessica's autonomy was central to the show's allure. For these women, their self-identifications with aging and with their gender positions were deeply intertwined. Their goal of sustained independence could not be linked with one issue without the other. They conveyed a growing sense of social vulnerability as women because of their mounting age. If they were to be strong elders, they would be strong elder women. This may have been especially significant for the middle-class women, whose autonomy seemed at greatest peril now and perhaps had always been. Some of these middle-class women told me stories about submerging their own unhappiness while subjugating themselves to fathers and husbands throughout their lives; now, they found themselves gently resisting subjugation to their children. For the less affluent women, who could depend on little support—or control—coming from their children, independence was theirs, like it or not; in fact, many of these women had adult children who depended on them for some form of support, often in the form of child care. For the most affluent women, loss of personal independence in advancing years meant moving to a nice assisted-living facility where amenities and diversion were provided; it meant no longer having to administer a large house. Neither the richer nor the poorer women welcomed loss of control, but such losses seemed mitigated by other factors. As I was told many times by the women who were in the middle economically, the threat of dependence meant losing the car keys and perhaps the ranch house. It meant, upon entering widowhood, learning to read financial papers and trying to make judgments about how well one could afford to live. It meant moving to a place where one did not want to go and turning over one's modest resources to the stewardship of one's children.

Acceptance of sickness could mean financial catastrophe. It was only logical that Jessica Fletcher's little manless victories would carry such symbolic weight for them.

The more working-class fans tended to be drawn by Jessica Fletcher's inner qualities. It was on this level that they felt they could most readily identify with the character, because the values and belief system that she seemed to represent resonated with their views of their own lives. For example, Naomi, who, like several other members of the Northside group, had moved to Milwaukee from the rural South as a young woman, said: "I can identify with her, because she's intelligent and well dressed and still has that air of small town about her."

Another Northside resident, Mary, said, "I just like the way she handles herself," with Naomi adding, "She does a good job with everything she does. She's a morally nice person, that's why." Hattie, another Northside group member, noted that Jessica was "truthful in a Christian way." Justine elaborated: "I feel that the only way that Murder, She Wrote comes through these situations that she comes through [is] that she's got God with her. She doesn't have to use a weapon or show a weapon or nothing, so she has got to get it from up there."

With only a couple of exceptions, religion is mentioned only ceremonially in the course of the *Murder, She Wrote* narrative, with Jessica showing up at church weddings and funerals. At no time does the character pray or otherwise seek God's help. Yet the Northside women "read" the program as a Christian text. Even if Jessica does not explicitly invoke Christianity, they perceived her to lead the Christian life. These women, for whom prayer, faith, and holiness played a significant role in everyday life, perceived Jessica's success to be rooted in her apparent Christianity—a state of grace. This starkly contrasted with the higher-status women's full crediting of Jessica's blend of intuition and logic. The character's humility seemed of paramount importance to the Northsiders. Several times, they complimented the character's attempts to improve herself over the years; crucially, they noticed, she remained "good" despite her success. They said they all found strength in Jessica's unflappable nature, despite the stress and storms that plagued her. They had endured enormous stress all of their lives, being plagued by continuous economic uncertainty, family pressures, and lack of social recognition. They lived in a high-crime neighborhood where some of them knew contemporaries who were crime victims. These women knew what it meant to adjust to life's setbacks. They, like Jessica, knew that keeping a cool head and taking the next logical step was the pattern for "success."

Program Content

Almost all the fans said they liked *Murder, She Wrote* because it was "a good, clean program," "not violent," and "free of rough language." This "clean" program concept seemed very important to them; they reported themselves as "selective" TV viewers who avoided much of network programming because they found its content offensive. They said they found *Murder, She Wrote* more as they fondly remembered television—and movies—to be when they were younger.

Among the upper-middle-class women, Helen, of the Christian home, went as far as to say that *Murder, She Wrote* featured a "nicer class of people" than even comparable programs:

> Years ago, I knew a lady who was so proper and so religious. But she watched soap operas. And I said, "How do you relate to this, because you don't believe at all that stuff that goes on? How can you do that, the rape, the murder, and all the kinds of stuff going on in the soap opera?" And she said, "It happens to such nice people." And that's what I think the difference is between *Murder, She Wrote* and *Matlock*. It's a different class of people. Such nice people.

In her own fan group, Lillian, a middle-class Jewish woman, got echoes of approval when she expressed her outrage at the offensiveness of most primetime dramas:

> I like to turn [*Murder, She Wrote*] on because I know she will not be taking off her clothes and jumping into bed with the nearest man who crosses her path. There are other programs about young detectives and young lawyers and young doctors. I have watched every one; they're all in their 20s and 30s. I don't care if they're intelligent, they're college educated, they have big careers. Somehow, ten minutes after the program starts, the dang fool woman is in bed with some stupid man that doesn't know his behind from a hole in the wall.

The conversation with the group of suburban women struck a similar chord:

Sharon: I'm not averse to saying, "Oh, damn!" when I hit my finger, but I don't really like for them to S-O-B all over the place. What does that prove? . . . It's a bad story, so you swear a lot and have sex—that makes it good?

Jane: The people in *Murder, She Wrote* may be just as dead, but they die in prettier positions.

Sharon: It's the mind solving the crime that's intriguing, not the murder.

Jane: The work. It wasn't the shock value.

Sharon: When they can solve it and get the guy that did it in one hour and put him in jail. I think that's neat.

Jane: Certainly beats O.J. Simpson.

Rural residents and Northsiders felt the same way. "It doesn't show killing," Naomi explained. "Even though that's what the story's about, you read between the lines. I like her because there's not a lot of violence in it. It's something I'm used to seeing years ago, an old-fashioned murder mystery." Some of Naomi's fellow group members said they would have preferred their young relatives to watch *Murder, She Wrote* instead of the more violent programs on television, but few did. "It would show them that without violence you can do things in a sensible way, without getting ugly," Mary said.

That the fans' pleasure in the program"s "clean" quality crossed class and ethnic lines gives support to the idea that elders generally are more socially conservative than are their juniors.[39] As I argue elsewhere in this book, this social conservatism may well be less a function of age than of cohort: the women in this study came of age well before the period of so-called sexual liberation and tended to hold more or less traditional notions of "decency" as the concept relates to media portrayals of sex and violence. In fact, *Murder, She Wrote*'s purity aside, elders I have met generally have commented on their disgust over television's gratuitous sex and violence. These concerns resonate with broader middle-class anxieties about the exposure of children to sex and violence on television.[40] While elders often have expressed worries to me over a perceived third-person effect as they gauged children's vulnerability to television's charms, they just as often told me that they found sexual and violent content personally offensive. It represented the seamy side of American life. Televisual sex, particularly, offended them because it represented a vulgar elision between private and public aspects of life. These people had lived as sexual beings themselves but belonged to cohorts that had come of age sexually in a context that insisted sex be practiced not only privately but secretly. Similarly, the more conservative of these elders felt repulsion when they encountered popular themes dealing with incest and domestic abuse: these topics had been taboo for most of their lives, and it felt discomfiting to have them explored in a medium that too often now publicly articulated private worries.

Beyond *Murder She Wrote*'s perceived "clean" quality, the upper-middle-class group members complimented the program's charming sophistication, which they tied to "harmonious theme music," the "consistently good quality of the writing," the "believability of the plot process," and the "inventive" combination of settings.

These fans did not associate "quality" writing with realism. I asked the women at the Christian group whether they found the writing believable:

Donna: You mean as a rule they always pick the wrong person, and after a while she solves it differently. Well, we are accustomed to that. We know it is going to happen.

Bea: I accept this story by itself. It doesn't matter to me what happened the other times.

Most of the fans received the show's coincidences with a sense of humor, as in this passage from a meeting with the suburban group:

Sharon: Wherever she goes, there's a murder.

Jane: She leaves Cabot Cove, because if not, they would have killed them all. The whole population.

Sharon: Nobody left but Jessica.

One upper-middle-class woman I met, 77-year-old Jackie, said she liked watching "the old *Perry Mason* reruns" on cable channels for similar reasons:

> Of course, one lawyer would never just happen to turn up when so many dead bodies are discovered. But that's part of the fun of the program. I mean, you can count on Perry going through every step of the way—uncovering each person's motive, maybe making one person look guilty when it turns out at the last minute they have an alibi, and then, finally, that

courtroom scene, when he catches them and makes them confess. It works out clean every time.

The lower-middle-class women I interviewed also made observations about the production of the show. Alma, for instance, noted, "Last night, I saw a little thing laying down there [in the shot]. I don't know what that thing was, either, but the camera focused on it. It was going to be important." She added: "The friend who is accused of the murder is always innocent. The format makes it easier to follow along." Her fellow group member Paulette, who felt that the show was "more carefully crafted than most," observed: "She's got a format that runs: introduce people and the problem and then comes some more problems and then it's resolved. That's what I like about it."

For all the show's incredibility, most fans called it realistic. It either recalled a special time of life, as it did for some of the African American fans who had moved to Milwaukee from the small-town South, or it represented an ideal that was possible in the present. As suburbanite Jane noted, "When I was growing up in Chicago, I would have told you that Crabapple Cove doesn't exist. But over the years, living in Wisconsin, I quite believe it does." This nostalgia for an ideal past conveyed the concerns of all these women—across lines of class and ethnicity—that the world had grown too complex and even self-destructive. They linked *Murder, She Wrote* to a feeling of "the good old days"—a feeling edited to filter out the memory of inadequate resources, from utilities to medical care. Some of these women explained that it was not that they rejected living in the present but they regretted that the present seemed so complicated compared with "simpler times." When Jessica fished on a friend's trawler, then whisked off to Monaco on a jet, that told them that small-town, old-fashioned ways could be preserved even in an atmosphere of modern convenience.

Viewing Experience

Despite their protestations that coincidences among episodes did not bother them, the upper-middle-class fans did not seem to count episodes as singular experiences but liked to recover a familiar feeling with each installment. Several times over, they reported that they did not recall episodes and, in many cases, recurring minor characters. Sometimes, when I mentioned plotlines, some fans recalled them; the only plot they volunteered from memory was the Christmas 1993 show that stood out for them because there had been no murder, a break from the program's firmest convention. (No one volunteered knowledge of the episode's year.) Helen and Bea noted that, in their 80s, their memory capacity was not as good as it once was and that they did not mind watching reruns of the show because they had forgotten "whodunit."[41] They said they liked the show's tidiness, its self-contained plot structure that never leaves any detail left unresolved for the next episode. As Jackie had noted, it was important to all these women that all the puzzle pieces appeared sequentially—that the plot "worked out clean every time."

As in the case of most others, the upper-middle-class women all said they had looked forward to seeing the show appear each Sunday evening after *60 Minutes*. "I feel comfortable, you know," said Analise. "I watch *60 Minutes* and I know, here she comes. . . . I went to Germany for a visit and I was so homesick. I turned on the television and she was on, Jessica, and I was at home in America!"

Most of the women told stories about ritual experiences over their contact with the series. For many, this had been a private time, as suggested earlier in the chapter, but for a few women, such as suburbanites Jane and Sharon, Sunday evening had meant a "date" with one's spouse to watch "Jessica"; their husbands liked the game of competing against them to see who could figure out the killer's identity first. (The husbands were unwilling to be interviewed because they considered *Murder, She Wrote* to be a "women's show.") Sharon and her husband had gone as far as to tape the program if they had made plans to be out. "We'd watch it another night when there was nothing on," she said. One other woman, Cynthia, the lesbian elder I met, said she had taped the program when she had to work at her job as a hospital nurse on Sunday evenings and, during the show's last season, on Thursdays. But the rest of the women either did not own a VCR or did not use it for time shifting.

One of the sources of that comfortable feeling for the upper-middle-class women was the parade of old film and television actors as guest stars. The dependable appearance of a recognizable fading star distinguishes *Murder, She Wrote* from many other mystery shows, which do not rely as heavily on this technique. Jackie compared *Murder, She Wrote* to the 1970s and 1980s series *The Love Boat* and *Fantasy Island*, shows that featured ensembles of recognizable guest stars: "My husband and I used to watch to see stars we *used* to see, and talk about how they'd changed or how we hadn't seen them around. Sometimes I find this is true, to some degree, with *Murder, She Wrote*. Somebody is always turning up there." Viewers such as Jackie might have remembered an earlier, pleasant time that coincided with "knowing" these actors, or, as Patricia Mellencamp has observed about these *Murder, She Wrote* star sightings, "We measure these many guests against the last time we saw them."[42]

The *Murder, She Wrote* fans did not speak of comparing themselves to the actors' appearances but talked about the good feeling they got from seeing these people still vital in their old age and, as Mabel described it, enjoying Lansbury's "charitable" provision of employment for them. The lower-middle-class fans spoke similarly: "It's nice to know they're alive," Fran said.

One of the things that people said made the show feel "comfortable" to them was the regular opportunity to compete or join with Jessica in determining the identity of the killer. Most of the upper-middle-class women said they liked participating in each episode's puzzle-solving process; generally, they said they guessed wrong but it did not matter to them.

The less affluent women of Milwaukee's north side tended to say that they liked trying to figure out the killer's identity, and, like the more affluent women, it did not bother them that they often guessed wrong. Naomi told about her experience with *Murder, She Wrote* and its genre: "I like the mystery shows. They're full of anticipation. It keeps you wondering what's going to happen. It makes you think. It's the drum music and all the music that goes with it. You kind of know to anticipate what's going to happen and all the time they keep you sitting on the edge. If it's going to be something nice, then the music's soft, and if it's not, well . . ."

Women from all the groups tended to say that sometimes all the clues they needed to guess the killer were available but that other times pertinent information would be hidden until the pivotal scene when Jessica would announce the solution. As suburbanite Sharon noted, "Sometimes she throws you a curve."

Murder, She Wrote's emphasis on episodic conclusion was essential to the elders' enjoyment. The Northside viewers said if an episode continued into a second week they might have trouble picking up the story line or worry that they will get busy with something else and forget to watch the show.

At the time of the interviews, CBS either was about to move the show to

Thursdays or had already done so. The program had been the clear favorite of all the women interviewed, but about half seemed confused about when the program was then in the weekly television schedule. Many of the women, including some who had successfully been able to follow the program to Thursday nights, expressed indignation. Some took it as an affront by the network against Lansbury, in whom they saw their entire age cohort symbolized. "I don't blame her for being mad," they would say. Others, generally the more affluent and better-educated fans, said they realized the move was about demographics. "It makes me angry that the older audience doesn't count," one woman said.

The Northside women said they would miss Jessica's visits to their living rooms. For most of them, these had been a premium point in the week, and some had even refused to answer the telephone when the program was on. "They know not to call me," said one of the women. If friends dropped in during *Murder, She Wrote*, these women said they would answer the door, but most probably would leave the set on.

Coming Out Clean: The Mystery Solution

Numerous differences emerge between the words of the various groups in this chapter, although in many cases these are differences by degree, not differences of a qualitative nature. I believe that the most important one is the contrast between the more affluent white women's tendency to project self-identification onto the star/character in such abstract terms as the search for personal autonomy and the lower-income African American women's more urgent tendency to project onto the character their own anxieties about crime and youth in their urban setting and, in some cases, their spiritual worldview. Such differences point to these women elders' initiative to view television productively and creatively.

These diverse elders approached their favorite program uniformly on many fronts, such as the rightness of an older woman's intelligent involvement in society and the desirability of television shows that do not offend them. But the more diverse readings—about Jessica's dealings with police, men, and youth, about the relative importance of Jessica's assertiveness or her compassion for those less fortunate—confirm these fans' willingness to produce meanings from television that are serviceable for their own purposes. Indeed, it is heartening to see some of these women establishing their power with respect to controlling the television appliance: sending their husbands to other rooms to watch an inferior set when *Murder, She Wrote* was on, refusing the interruptions of telephone callers, taping the show to avoid missing it when they had to be out for the evening. Hardly any other television program, these women felt, warranted such a controlling response on their part. This told me that *Murder, She Wrote* provided something for them that they did not—or could not—get anywhere else in their lives as they currently were constructed. (I spoke with a few of the women about a year after the series ended, and they still felt a void in their life. They *missed* connecting with Jessica and feared that the rare combination she represented for them—an attractive, independent older woman working out problems clean in an entertaining show without graphic sex or violence—would always be missed.)

It remains obvious enough, however, that *Murder, She Wrote* conservatively portrayed comfortable social messages that helped these fans avoid thinking about alternatives to their own views. These were women who refused to watch *Roseanne* because they found the star "revolting" and "vulgar" and who disliked *Law and Order* because the bad guys frequently got away; indeed, it was difficult sometimes to discern who the bad guys were. *Murder, She Wrote*, for them, was brilliant simplicity with homespun moral preservation. Unchanging, resonating ideals were plain to observe in the text, and, if a fan did not wish to probe deeper matters than Jessica's hemline, she was not forced to do so. The formula, however one might read it, was satisfyingly static and therapeutic, whether the issue was about reassurance concerning the innocuous nature of technologies that might, as fans first learned of them, appear frightening or about the unfailing reinstallation of order when the puzzle was solved.

Three

Television Use in a Retirement Community

After her husband died and Barbara retired from her internal medicine practice, she was finally ready to attack her reading pile. "I had accumulated all this reading I was finally going to have time to do," she said. "It's ironic, really. I thought I would just read up a storm, and I found I couldn't do it. My eyesight had grown dim, and I found I just couldn't *cerebrate* like I used to. I found television, a visual and auditory medium, easier to digest."

Television became Barbara's focus. Eight years after she became a widow, she left her small-town neighborhood and relocated to Woodglen, a retirement community in a nearby university city in the Midwest. Instead of replacing television with organized Woodglen activities, she began making it central to her life. She has become a cable information addict, in particular, a self-described C-SPAN junkie: "I see now, I didn't know anything about anything before. The whole world has opened up to me." In her mid-70s, Barbara has joined political causes, particularly environmentalism, whose agendas she was exposed to on C-SPAN. She writes to legislators. She shares information with her neighbors.

Television has superseded part of Barbara's need for human companionship. While she has partially integrated herself into her new community, participating actively in communal dinnertime table talk, she avoids most organized activity and spends the majority of her days and nights alone in her garden apartment. Especially when chilly weather prevents her from working in her flower beds, she spends hours in her wingback chair, connecting shreds of cable content with the help of printed viewing guides and two remote controls that she keeps in the adjacent table drawer.

She has, over the course of a typical day, "witnessed" hearings, speeches, and roundtable discussions on C-SPAN and picked up supplementary news and commentary on CNN. When it is time for the evening meal, Barbara puts on presentable slacks and drives the half mile to the main building at Woodglen, where she joins acquaintances and perhaps her cluster of intimates, whom she has converted to "C-SPAN junkie" status, at a table for six or eight. Conversation quickly turns to the substance of television news and C-SPAN. The talk about the station's goings-on coalesces into a sharing of ideas on timely political issues, which may include health care reform or abortion rights.

At dinner's end, Barbara drives home and watches PBS's *MacNeil-Lehrer News Hour*, the C-SPAN call-in show, and other cable programming, such as *Larry King Live*. All of this has been plotted carefully, with adjustments being made according to new content she learns about at dinner. The material she gets from these telecasts may inform her position at the next roundtable dinner, but she will assimilate it into what she already knows and what she knows of what her neighbors think.

Barbara is one of tens of thousands of Americans who live in the nation's growing numbers of retirement communities. Like many others, she has found an important place in her days for mass media, particularly television. Not only has she come to rely on it to fill new spaces created by the loss of past associations, but she has discovered novel applications for television as well. Along with her neighbors, she seeks out content in order to learn, to be entertained in ways that are new to her, and, perhaps most significant to her, to arm herself with information that grants her membership in the social network of her community. Far from the image of the dulled elderly person who sits in front of the television set without direction, Barbara and people like her have chosen television with purpose. In doing so, they are carving out some new meanings for it in their lives that the television industry may not know much about. Mass media scholarship traditionally depicts older viewers as eager to accept whatever fare is available, with a preference for news and mystery genres.[1] These depictions are only partially accurate for the viewers at Woodglen. Considering this community's use of television may spur us to reconsider how the elderly infuse their lives with meaning, in part by incorporating media into their daily lives.

In the case of a retirement community, where most residents live alone and share many basic everyday experiences, such as some of their meals and transportation, study of the household as the basic domestic unit of media consumption is too limited.[2] In the Woodglen retirement community, two spheres share significance, the domestic and communal. These spheres—and what is taking place in the spaces between them—are where we can look for answers to questions about how Woodglen people make meaning from media.[3]

Through initial volunteer sign-up and informal networking, I was able to meet with 26 of Woodglen's 290 residents. My sample reflected a more or less balanced mix of the community's residents: about three-fourths were women, about two-thirds were widowed (with one divorcée), and about half had a background in academics, either personally or by marriage, and the other half in some other aspect of professional or business work. About a third of the women had been homemakers. Over a period of two years, I interviewed residents singly and in small groups, conducted focus groups, and observed residents in their homes, in front of their televisions, and in informal social gatherings in the community. It struck me that, while Barbara may have watched much more television than many of her neighbors and relied on it a great deal more both for companionship and to oil the machine of social intercourse, almost all the Woodglenners both hated television and loved it. They derided it as a medium and depended on it, to one degree or another, for many of the same reasons that Barbara did.

The Woodglen Community and Television

Woodglen is located on the fringe of a large university campus. The community, once formally connected with the university's music school, was populated exclusively by retired academics and their spouses in its early years. The property was sold to a for-profit corporation, and the academics' majority is diminishing. Three circles of duplex apartments on wooded, landscaped grounds form the majority of the dwellings. Scaled-down collections of the residents' high-quality furnishings fill these flats, and even smaller collections go into the apartments in the multistory main building. Off to one side of the main building is the convalescent wing, for people who can no longer live independently. The population is homogeneous. Exclusively white and upper middle class, Woodglen's inhabitants are college-educated, with many holding graduate degrees. Political views are mainstream; Republicans and Democrats are plentiful, and the occasional Independent is found. Protestants are in the majority; there are a number of Catholics and Jews. Ranging in age from their mid-60s to 90s, residents experience varying states of health and maneuverability.

The main building's communal dining room, the heartbeat of the community, serves two meals a day, lunch and dinner. At lunchtime, meals are served in a cafeteria line, and residents, unless they are going through the line with friends, carry a tray to a table and are seated "potluck." At dinnertime, they may still be seated, this time by the hostess, at a table with others unplanned, or they may arrive with friends and be seated together. At this meal, a server takes orders from a short menu and brings courses one at a time. Sport jackets and dresses are the norm at dinnertime, and the occasional fur appears amid wool coats in the cloakroom. Sending a message about the community's "active" lifestyle, Woodglen prohibits wheelchairs in the dining room. Residents rarely drop in on one another without calling ahead, but they frequently enjoy encounters in quasi-public spaces. In-home entertaining tends to take place in luncheon or coffee contexts or occurs when a woman or couple invites a cluster of friends over for pre-dinner cocktails before the group adjourns to the dining hall. Committees, classes, and cultural events bring small groups together frequently. The atmosphere in these shared interest groups is cordial and agenda-oriented.

Inside the residents' homes, the use of furniture in the television viewing area tends to support Woodglen's status as a retirement community that emphasizes "active" living, and it also supports the atmosphere of formality that I discovered there. Although sofas and overstuffed chairs are common furnishings in the viewing areas, recliners are not. Some people even used straight-back chairs in the TV viewing area. A number of late-night viewers told me they generally turn their sets off before dressing for bed. Such observations led me to understand television viewing at Woodglen as a discrete, purposeful activity.

Residents of Woodglen tend to do most of their television watching alone inside their individual apartments, but some viewing occurs in shared settings. Encouraged by the easy flow of human traffic in a densely populated zone of retirees, people occasionally invite one another over to watch a tape. Several residents take advantage of frequent opportunities to view videotapes of films or live telecasts of college basketball games on the big-screen television in the main building's terrace room. Although these showings do not attract more than a dozen people, they represent a shared audience experience in the community.

More pervasive than shared viewing experiences are occasions within a person's day when communal life replaces experiences that formerly might have occurred within the household. Arlie Russell Hochschild, in her study of women living in a congregate facility, found such replacement of former, family-focused experiences by activities in the communal setting to be integral to the retirement living she observed.[4] At Woodglen, daily meals at tables for four or six in the dining room, conversations around the hearth in the sprawling communal drawing room, and similar moments bring small groups together informally, in places between the intimacy of home and the decorum of public space. Apartments feature quasi-public spaces where residents come into frequent contact with one another: shared corridors in the tower, glassed-in porches along tiny, unfenced yards, and shared mailbox facilities. The emphasis on communal space does not eliminate residents' sense of private space. People control the traffic into their homes, and their possessions are theirs alone. In the case of widows, most possessions are held over from marriage, with a common exception being the gear of new technology—VCRs, cordless telephones, and answering machines.

Community Discourse and the Reception of Television

Each household has a place in the retirement community, and residents' understanding of television depends on their movement between their homes and the communal sphere. The community structure encourages certain perspectives and ways of spending time. First, most people at Woodglen pride themselves on not "living in the past" but instead on focusing on the present. To them, every hour is important, even if many of them must spend much of the time resting. For this reason, the only acceptable television is that which is perceived to be relevant to the time in which it is viewed. (For example, old entertainment programs are perceived as irrelevant; historical documentaries written for today are perceived as relevant.) Second, the community collectively puts a high value on cultural items—

including television content—that appear to be of high quality. The communal bias that favors high culture over popular culture spurs residents toward what they perceive to be "quality" television. Third, Woodglen's residents understand their community to be "civil" and "dignified." Their pursuits must reflect activity that they perceive to be of consequence. Desirable television, for them, reflects this concern with how time is spent. These three themes represent discourse that shapes and is shaped by the community. Likewise, this discourse shapes and is shaped by individual contact with television: talk between neighbors is colored to a substantial degree by the residents' knowledge gleaned from mediated information, and residents make everyday decisions about their own media consumption based on their neighborhood talk. For most Woodglenners, decisions about media are decisions about television.

Living for Today, Not for Television

Some Woodglen people talked enthusiastically to me about television; others disliked it. But when they talked about their feelings toward it, they spoke in terms of currently produced content, not in terms of television's entry into their media environment four or five decades earlier. For example, most subjects said they saw "no purpose" in watching reruns. They said they never watched old television programs such as *The Lawrence Welk Show*, domestic family situation comedies shown on Nickelodeon's *Nick at Nite*, or the American Movie Classics channel. They preferred new programs and were selective about genres and channels. (Many avoided popular entertainment altogether and watched only nonfiction forms.) As one woman, Fostine, explained, "I don't live in the past. I live for today."

Many residents told me they thought most viewers of television, the aged included, either condemn television for its negative qualities or indulge in watching worthless content. A friend of Barbara's, Daphne, told me she felt that she was able to make critical distinctions about wide-ranging content, although she felt many others were unable to do so and were vulnerable to content. "I am 70 years old," she said. "The world is changing. I cannot force my views on other people. I want to go out and take people off the street and sit down with them and say, '*There's* responsible TV. It tells you what you need to know *now*.' "

Fostine, in her 80s, told me she has watched about an hour of television a day since moving to Woodglen. Her comments were comparable to those made by several widows:

> After my husband died, though, which was in '84, I found myself getting into *terrible* habits of sitting down and watching TV all day long. Good, bad, didn't matter. Half the time, I didn't listen. I might have been reading at the same time, but it was on, probably because it's company to have a noise in the house instead of having it deathly quiet. But now that I've moved here, I'm so much busier, and I'm not lonely, and I don't watch much television. Now, I watch to *watch* when there's something I want to *see*.

Network and CNN news programs, *Jeopardy*, and PBS's musical offerings are generally the choices she makes.

The people I met at Woodglen possessed the academic and cultural capital that propels many of them toward the arts. For instance, some of them decorate their apartments with expensively framed and matted prints of painters they consider to be important and art objects that they acquired from visits to foreign lands. Pierre Bourdieu, in his critique of the relationship between aesthetic taste and what he has labeled cultural and economic capital, perceives television as a carrier of the "popular aesthetic."[5] He concludes that when television "brings certain performances of 'high' art into the home" for a working-class public, it creates "what are virtually experimental situations."[6]

What has occurred at Woodglen differs from the experimental situations Bourdieu has identified: television, conventionally a carrier of the popular aesthetic, is bringing performances of high art and other elite cultural forms (e.g., much of C-SPAN's programming) to an audience that has the tools to consume it skillfully. There is a great amount of talk among informal groups about the elite contents of PBS—who saw the opera adaptation last evening and how they evaluated it, for example. Residents who said they might not seek out such content on their own said they felt some pressure to watch it in order to succeed socially at Woodglen. For others, television offers a limited means to keep up with activities that they do not want to give up as circumstances change.

Ironically, most people I met at Woodglen saw in their consumption of television a tendency to love what they hate. For example, Don, a man in his 70s, volunteered, "Even though I think that, on the whole, the quality of television is very poor, I have found some things that are quite worthwhile, with good writing or with interesting information. I use it to keep the learning process going." Television has helped Woodglen residents engage in the retirement lifestyle they expected to enjoy. Like Barbara, many Woodglenners moved to the development expecting to expand their involvement in activities they loved—only to find, as Barbara did, that their aging had preempted many possibilities. Most of them found television's "lifelong learning" promise—or its vicarious benefits of such desirable activities as travel and exposure to high-culture events—an acceptable substitute, as Don did. That television offered to replace many important elements of their retirement dream helped them stave off grim thoughts of life's essentially being used up. "I can't play golf and I can't travel," Don told me. "I'm not able to just take off and fish any time. I really don't just want to sit here." With the television set turned on and tuned in to something that he could learn from *today*, Don felt that his Woodglen lifestyle was somehow assured.

The Search for Quality Television

Woodglen illustrates how a community can help to shape personal intentions toward television. This illustration suggests a complicated relationship between these older viewers and the programs they watch. What has been suggested by Fostine, and several widows like her, is the possibility that particular retirement community living can help promote a *different* experience with television, not just less frequent contact. These widows no longer turn on the TV to create a "noise in the house." They turn it on to get information they wish to have. While their particular backgrounds, including college degrees and the life experiences that often went with them, no doubt encourage their desire and ability to consume knowledge-oriented programs, in case after case, the move to Woodglen precipitated their turn to more such content. Most residents felt that their use of television was fairly consistent from the time they bought their first set until they retired. Almost invariably, the amount of television watched increased with such changing lifestyle factors as separation from children, retirement of self or spouse, retreat from an active social life, death of spouse, personal infirmity, and loss of autonomy in transportation.

One factor may separate the Woodglen people from some populations of aging Americans who increase their television viewing. Most of them have met retirement with a wish to remain productive and serious, by their own admission, and this desire carries over into their television use. Residents frequently come together informally, as dinner companions, cocktail hour participants, and weekly coffee hour goers. In these situations, talk inevitably is about what people know and are interested in. Most often, this talk is about people's knowledge of current events and their impact on the world, knowledge of faraway places in which they have traveled and perhaps worked, information about a subject to which they have had some special exposure, or particular career experiences. This purposeful mode of conversation encourages talk about some television forms while precluding discussion of others. It was common for me to hear people making enthusiastic references to PBS's musical programming. On the other hand, talk of fictional genres, soap operas particularly, is shunned at Woodglen. Most people I talked with are generally unfamiliar with network lineups beyond a handful of the most popular programs among older people more generally. (Although members of this crowd knew of and had perhaps seen *Diagnosis Murder, In the Heat of the Night*, and *Matlock*, most were not regular viewers. These shows, although they generally attracted a large following of elderly viewers, were considered pedestrian by most Woodglenners.)

Recent situation comedies were almost unknown to my Woodglen respondents. Although none of the people I interviewed was a regular viewer of *Murphy Brown*, which peaked in popularity during my involvement with the community, a number of women mentioned the program's lead character as a positive indicator of women's advancing position in society. "Women do so much more than they did years ago," Fostine told me in our first interview in her home. "Often, television shows reflect that, as I understand *Murphy Brown* has done." A few months later, during the course of the 1992 presidential campaign, Vice President Dan Quayle used the program as a vehicle to criticize single motherhood. Although Fostine continued to think that television could show positive role models, her view of this program changed. She had held a career and supported women's success, but she decided that the liberal Murphy Brown, by betraying "traditional values," had contributed to television's "bad" content.

Among people who watched fictional programs infrequently, quality was the crucial consideration. This translated into well-defined preferences for middle-brow programming, such as the content of PBS, the more respected portion of A&E's nonfiction texts (such as the *Biography* series, at least prior to the mid-1990s), and the like. Residents shunned any content that they identified as popular. As Barbara explained: "As my selections now are, rarely do I listen to a fictional type of thing. Occasionally, I will make note that a *Hallmark Hall of Fame Special* is going to be broadcast, and I will watch it. I know that it will be well produced, based on something that is well written, and it will be well performed."

Her reasons for watching the Hallmark specials are similar to reasons people gave for watching fictional programs on PBS even though they do not watch commercial entertainment programs. The absence of commercials and the ostensible high cultural value of these programs, most of which are British-produced, also contribute to their popularity at Woodglen. I believe that the attraction to such shows is, for many people, a mixture of both recognition of merit and pretension. The PBS programs, the Hallmark specials, and certain news programs can be characterized by an aesthetic element of "class" that attracts the Woodglen residents. The urbane people on these programs use language well, display critical thinking skills, approach events and issues with a degree of emotional distance, and otherwise signify affluence. This is familiar and comfortable to Woodglen people, who generally find country or rock video channels, televangelists, daytime talk shows, and home shopping networks discomfiting. As Charlotte Brunsdon has suggested, some people (males, elites) avoid such less reputed genres as soap operas, in part, because they lack the cultural competence necessary to decode and appreciate them.[7] The cultural competence of Woodglen residents directs them to feel at home as an elite audience, seeking out the contents afforded by the new cable and satellite technologies and finding part of their own heritage—a collection of texts that presents traditional discourse on public and cultural affairs. The social background of Woodglenners helped to constitute them as what Herbert Gans has identified as a "taste public," a group that exercises certain values with regard to cultural forms such as music, art, literature, drama, criticism, news, and the media in which these forms are expressed.[8] Taken together, the Woodglen "taste public" lies somewhere in the overlapping space between what Gans has identified as high culture (dominated by elites, including academics) and upper-middle taste culture (dominated by the professional and executive classes). Gans's work on taste cultures appeared more than twenty years ago, so it did not account for the current range of television offerings and the divergent status positions that such offerings might occupy.

Allowing for an updated reading of television, then, this overlapping high and upper-middle taste culture occupied by Woodglenners privileges the elite forms of television, such as *Masterpiece Theatre*, as well as what Woodglenners take to be "serious" nonfiction content, to be found on public and many cable channels.

Leisure as Work

In Western industrialized societies, the concept of leisure for the elderly, in contrast to work, is relatively new, a function of the economic productivity of this century. This new leisure tradition has been heavily associated with recreation, although *leisure* may have very different meanings in different communities.[9] In many Sun Belt retirement communities, leisure may be interpreted as recreation and is clearly associated with activity as contrasted to atrophy.

In the Woodglen environment, there is a certain amount of structured and unstructured frivolity. Wednesday-evening poker games take place in the communal lounge area, and a cash bar is opened every Friday evening for a social hour. Such events attract minorities of residents, however. Many people told me that they view their retirement years as a time to do new work or continue old projects. As Verna, a retired substitute teacher, noted: "People here stay young, because they are so busy. They come here to stay busy. Most of them have moved here because they need to get out of their big homes and get some help with things like cleaning and meals. Or they come here to be near children and carry on with life. But they don't want to give up their independence." For Verna, "leisure" means business: "I watch very little television. It has to be important, because I don't have time." She and her neighbors seem continually concerned with whether their time is being well invested. Opportunities to "invest" their leisure at Woodglen are a primary reason many find life there satisfying. High-culture activities are perceived as worthwhile investments, and community talk about them is rewarding for participants.

Watching television can be a form of leisure work for Woodglen residents, and it most often is. For instance, one couple, Geraldine and Sam, told me they enjoy learning from travelogues and foreign-language lessons on the public television station. Television viewing is a highly structured, planned activity with particular rewards and outcomes for Woodglen residents, including increased knowledge and cultural enrichment. Residents rely on high-culture content, such as telecasts of opera productions and governmental proceedings, to afford them material for meaningful exchange with their neighbors.

Woodglen residents are concerned with mastering the information and entertainment content that is available to them through the media. To varying degrees, they employ desk calendars, printed viewing guides, remote control devices, and videocassette recorders to help them use television proficiently. In most cases, people set aside particular segments of the day when they will and will not watch television. Some of them pencil programs to watch into their datebooks. Generally, they let structured activities of the community take precedence when these seem important or interesting to them.

Midge, 73, is a retired military hospital field director. She has remained single and moved to Woodglen from a nearby city several years after she retired to "be near people, which you need." Midge is the lone self-reported soap opera fan among my respondents, and she said she is sensitive to comments about her desire not to miss *As the World Turns.*

"Of course, they don't understand that I may be ironing or I may be writing a letter, but I am busy," she told me. "When I was taking the water exercise class last year, it used to cut into *As the World Turns*, and that annoyed me."

"But you went anyway," I said.

"I suppose I allow the curriculum to interrupt my soap if I consider the opportunities important, but I'm still annoyed."

I asked her what other soaps she watched.

"All the CBS ones, *The Young and the Restless*, *The Bold and the Beautiful*, and *As the World Turns*."

"But not *Guiding Light*?" I asked.

"I have to stop somewhere! I can't just sit all afternoon and watch the soaps. I have to draw the line at three o'clock and do something else."

I asked Midge why she watched the soaps. After telling me it was something she had done for decades, having been introduced to *As the World Turns* by her mother, she said the soaps are "pure entertainment." Then she added, "And I do learn from them, especially about developments in the medical field." She told me about a recent story line having to do with developments in the treatment of colon cancer. "The message was the importance of early treatment. This is something people need to know. I think soaps do a public service. They're pretty timely."

It seemed important for Midge to justify her time with soap operas, as she did with game shows. (She often watches *Jeopardy* with an infirm neighbor, who she said needs the "socialization.") Whether people bring this work-oriented attitude toward leisure with them to Woodglen or whether they learn this attitude in the community, it is unquestionably present, and it influences television use profoundly.

The viewing of televised sports constituted a special male-sphere activity for the women at Woodglen, and it was heavily influenced by the place of men there. As I have noted, men constituted only about a fourth of residents, and they were excluded from many communal activities by the very nature of these. For example, most of the men had little interest in ceramics, sewing, or flower arrangement, so they stayed away from many of the classes where the women congregated. Most of the men also refrained from involvement in social planning and weekly coffee klatches. Even if men had been genuinely interested in such activities, these were so heavily gender-coded that they probably would have felt embarrassed or intimidated by the prospect of joining in. Where the men generally did appear was at events that focused on high culture, such as bus trips to university cultural events and concerts on the Woodglen grounds; evening social gatherings, such as the weekly cocktail hour; and, of course, the daily communal meals. Men remained outnumbered in these parties, but, as a function of mortality statistics, they were outnumbered generally.

Jack, a widower in his 80s, appeared in the dining room one night and joined a table of single women with whom I was sitting. Immediately, they made a fuss over him, and he gently dominated the evening's conversation among the group. At one point, he turned to me and jokingly said, "I'm valuable here, you know. There aren't many of us." Catherine, who came to Woodglen after divorce ended her longtime marriage, added: "Most of the men here, in fact, are with their wives. Many of them stay to themselves or associate with other couples. We have a serious depletion of men with whom to converse."

Understandably then, perhaps, the rare available single man such as Jack often was treated deferentially by women when it came time to set a topic for discussion. These same women had been used all their lives to deferring to the wishes of men, and now they did so in order to reward them for their company. For Jack and others, sporting events often crept into conversations with women, who almost never discussed these things among themselves when men were absent, except for the most rudimentary mention of the university basketball team's recent performance. A few exceptions occurred, to be sure; one or two women at Woodglen were known to be "fanatics" about the local team. But most of the women seemed to have begun earnestly to follow the team—and sports generally—since moving to Woodglen and living among people such as Jack.

Little detailed talk of sports occurred in the communal spaces of Woodglen beyond discussion of the basketball team. In fact, many of the men themselves had little interest in sports. For instance, Frank, a philosopher, once explained to me the pressure he felt to conform to the pressures of male identification with sports: "I have no interest in it, really. I worked here at the university many years ago, and I deeply identify with the university, but I do not care much for sports and never have. I have never watched them on television. However, I realize the significance of the basketball team here and have made a modest effort to keep up with it, at least through the newspapers, so that I can participate in discussion about it when I am called upon to do so." Generally, it was men who expected to carry on a conversation with Frank about college basketball; his wife, Charlotte, saw no need to learn about the team.

Many of the women Woodglenners took on their sports fan roles in earnest. Midge, for example, did not know a foul from a no-look pass, but she religiously wore the team's colors to the dining hall on all game days, as did many women and men alike. Most residents tended to watch the games on television, generally alone or with a spouse in their apartments. Woodglen management played the games on the big-screen television in the main building, but no one generally watched there. Knowing who won, wearing the team colors, and having some insight into the particulars of the game—such as the flamboyant actions of the coach or the cost of some controversial call—made the neophyte female sports fans feel knowledgeable about an arena that was worthy of discussion in mixed company at Woodglen. In these discussions, the men often leveraged for dominance, introducing their finer knowledge of the game and its history, which the women generally lacked. Basketball was a male-sphere arena. The women knew it, but they still wanted to play.

And it was a form of play in which they engaged, one of the very limited range of nonserious leisure areas that was sanctioned as a legitimate component of Woodglen talk. Sports was fun, even silly, some of the women felt, but everybody needed a bit of levity, even in a serious community such as Woodglen. These women's upper-middle-class background, however, precluded traditional forms of feminine levity; gossip would have been unbecoming and boorish. Their interest in the male-sphere levity of sports kept them from feeling that they were "working" all the time in their interactions with one another. Still, it was work. I had the impression that much of what the women knew about sports had been gained in rehearsal for position among their peers at the dinner table—women and men alike. And, although Woodglenners spoke of their home team with enthusiasm, they often did so unsmilingly. "What do you think of the coach's strategy to leave So-and-So on the bench?" one would ask, fully expecting an analytical reply from a neighbor. Animated it was not.

Because of television's social significance at Woodglen, people tend to look upon their use of it as interaction rather than as exposure. For example, Barbara said she considered her time with television a social exchange. Being widowed or otherwise left alone has contributed to such elders' drive to seek contact through television. Barbara's sister-in-law, who lives in another state, protested the Woodglen retiree's "excessive" use of television. Barbara explained to me: "She has missed the point, really. When she says there's no interaction, I can't argue with her in a way that she's going to understand that there is, to a degree, a certain interaction that goes on, and it isn't that I'm sitting here, you know, spinelessly, having the palaver on. I'm responding in my own mind whether I agree or don't agree with what's being said or I'm having the total picture or that sort of thing." While Woodglen has replaced Barbara's old neighborhood, whose households of working families were empty by day and busy each evening, she actively befriends "the tube." She admits searching out faces and voices to help her fill her days: "There are particular personalities I feel I interact with, in a way, by listening to them and thinking about the issues they raise." One of these is C-SPAN's founder, Brian Lamb, noted for his stony demeanor. Lamb often appears on camera to introduce segments, interview personalities, and serve as host of Sunday's *Book Notes* show, which is popular among Woodglen residents.

In Barbara's experience, as in others', she feels pressure to apply her time substantively. She and her neighbors feel they can use television to do what they perceive to be work in the context of leisure. This "leisure" work takes place in the public or social sphere as well as in the private or personal sphere as they use television in their talk with others.

The Woodglen neighbors, on the whole, share the philosophy that a productive retirement is a happy one. They reject the notion that, on leaving their careers, they left their identities behind and became more like retirees in general than like the people they used to know. It seems that many people at Woodglen consider their neighbors colleagues in retirement rather than members of a surrogate family. "We haven't reinvented ourselves," said Catherine, 80. She added: "Living here for me, and for most people, I think, involves pleasant interaction with others, which we all need. But the substance of my life is tied to the relationships I formed in my career. The environment here is quite civil, but not terribly intimate. It does help one stay involved with life."

Television and Public Life

The drive to use televison to "stay involved with life" came through most vibrantly in the story that Lila told me. Lila had only recently come to politics as an avocation when we met. Until a few years earlier, her life had been too full—a mix of public and private sphere activity—to devote time to what she considered democratic participation. She had been a biomedical researcher at a large midwestern university, one of the world's foremost research institutions. She and her husband had lived for a time in his home country, on another continent, where they both had been involved in research and had worked toward raising their family. An avowed rural dweller, she had long shunned the notion of communal living in her senior years. Then she found herself in a situation that has become all too familiar for many American women: her husband died, her offspring lived far away, and her friends, who remained "couples," liked to associate at night, when she preferred to be locked up in the cozy environs of home. Woodglen's conveniences beckoned to her, and she decided to "come home" to the community, having grown up in an adjacent county. In her mid-60s, she rented a two-bedroom garden flat and found that she could comfortably blend communal dwelling with an easy ability to drive to any outside amenity she wanted to reach. Quickly, Lila connected with adventurous neighbors—people, she said, who were willing to talk about ideas and get involved. She and many of her mostly widowed friends either developed a latent interest in national politics or strengthened a lifelong concern with it. Many times, this new level of political commitment both affirmed and bore the influences of new friendship circles at Woodglen. "Subgroups do emerge here," she told me. "You find them, as I did, in painting class and other activities." Such subgroups, Lila felt, were partly a function of political leanings and level of interest but also were influenced by such factors as residents' varying physiological capabilities: people who still drove and enjoyed going to see plays or movies, for example, understandably developed a chumminess that they did not share with others.

Lila emerged as a "successful" member of the community, because she tried out a great diversity of activities and enjoyed most of them. Because she was both a joiner and a worker, actively pursuing the tasks of committee work, Lila gained wide admiration at Woodglen. Her politics were strictly liberal Democrat, and this is the circle in which she formed her cluster of intimates, but she associated with almost anyone in the community who generally acted pleasant. Still, on a given day, if one were to encounter Lila at lunch in the dining hall with a friend, the conversation likely would be about, as she and her friends liked to say, "the issues." These talks resonated with the way national debates were framed by a blend of elite media outlets, such as the *New York Times*, National Public Radio, and C-SPAN. The talks constituted intensive, well-informed, and liberal-leaning discussions of, say, whether the Supreme Court might be headed toward a reversal of *Roe v. Wade* or how U.S. foreign policy in Eastern Europe was leading the country toward a dangerous mess.

For some of the people I talked with at Woodglen, public activism ended with community discussion and its pinnacle, voting. For others, such as Lila, it extended to community participation within traditional boundaries: joining the League of Women Voters, for example. Lila told me that C-SPAN prompted her to "try to help shape the public agenda by taking some responsibility at the grassroots level." She said she wanted to help get candidates to answer to citizens through her organization's work in holding public forums and conducting voter registration drives. Others, such as Barbara, did not join the League of Women Voters but did become avid letter writers to elected officials in Washington, or they joined and sent money to organizations that represented themselves as powerful citizen lobbies, such as

Greenpeace or Common Cause. In fact, Barbara briefly worked for a political campaign at Woodglen:

> I saw a little clip in the paper, placed by the local [Ross] Perot volunteer coordinator, inviting people to solicit fifty signatures each in order to help get his name on the ballot in the state. . . . I had no trouble in the dining hall. I didn't present myself as committing myself to Perot but asked people if they felt, as I do, that we should have as wide a choice as possible. Most people were very receptive. Those people who were indignant were all Bush people. They were all fortified Bush people, I should say.[10]

Lila's best friend at Woodglen was Emily. The two found, upon meeting there, that they shared some similarities in background as well as in interests, and they got together often to socialize. Unlike Lila, Emily had been a homemaker and had dropped out of college to rear her family. Her husband had been a professional, and she had enjoyed a comfortable upper-middle-class lifestyle that had emphasized both travel and continued education. She had always been interested in political and social issues.

Alone and with other friends, the two often talked about "the issues" as refracted through the lenses of the media outlets they monitored. Lila depended on newspaper editorials and letters to the editor in the state metropolitan daily for what she considered to be the substance of public thought. Most of her news came from public television's *MacNeil-Lehrer News Hour* (which later became *The News Hour with Jim Lehrer*) and C-SPAN coverage of proceedings that were relevant to U.S. government. Television was a less dominant medium for Emily, who remained an avid reader and faithfully toiled through the *New York Times*, whose size, she volunteered, intimidated her. She liked to process information from many media sources, with an emphasis on such programming as C-SPAN and *MacNeil-Lehrer*, in order to reconcile her position on issues.

Part of the attraction that C-SPAN held for Lila and Emily was its call-in segments, in which viewers raised questions and aired opinions. The viewers at Woodglen, these two women included, told me that although they liked to hear and later discuss the comments of callers, they did not call in themselves. They tended to cite the lack of an 800 number as a reason. They told me they found it more important to analyze what they considered to be a fair cross section of public opinion. They instead directed their personal views into interpersonal channels in the retirement community and in niches they had found outside it, such as civic or political organizations. The people who said they liked C-SPAN also said they liked *Larry King Live*, on the commercial network CNN; they appreciated the political significance of King's guests, his focused interview style, and the expressions and questions of "live" callers. A few people told me they did not favor C-SPAN because they found its coverage of events ponderous to wade through and they preferred the structured approach of more truncated traditional television coverage. This was a minority opinion, however. As Emily told me, "I enjoy the direct representation of the event." Lila agreed: "I hate soundbites! It's what I like about C-SPAN. You watch all of what's going on and make up your own mind." For Lila and Emily, *MacNeil-Lehrer*'s uninterrupted, fuller presentation of news issues—contrary to commercial networks' image-oriented and fast-moving coverage—similarly presented what they perceived as "all sides of an issue." This allowed them to arm themselves with what they considered to be raw data and commentary that encouraged autonomous decision making. For example, Emily said the *MacNeil-Lehrer* program gave "more complete details" than any other program did on the disintegration of Yugoslavia and the prospects for U.S. involvement in the matter. She felt that the program's coverage helped her sift through the issue so that she could decide how she felt about U.S. foreign policy. For Emily and Lila, more was better. They tried to produce their own agency, at least with respect to their understanding and framing of issues, using television as a tool in everyday life.

Lila's and Emily's use of such programming as C-SPAN began to constitute public participation in two important ways. The first way occurs when people such as Emily and Lila begin to experience political discourse less as "receivers" of media messages and more as people engaged in interaction. Their experience was limited to seeking texts, constructing sensory experiences, decoding messages, and reacting to the messages in their community environment, but they deserve credit for their activism. They reformulated issues as these were presented by the texts of media, especially television, constantly working them against what they already knew. They expressed enthusiasm about what they believed was available to them and considered these views in their actions as "citizens": in voting and working with the League of Women Voters, for example. Emily told me that she often compared what she read in the *New York Times* with what she saw unfolding on C-SPAN.

She said the "media perspective" often seemed the same from source to source but that callers on such programs as C-SPAN's viewer segment revealed positions she might not have considered.

A second way in which these Woodglen residents approached public participation was through their interpersonal connections, generally inside the communal settings of the neighborhood. When residents such as Lila and Emily introduced topics from their favorite television shows, as they frequently did, their discussion of the media representations reverberated with their personal reactions and perspectives. These were colored by the reactions they had engaged from their neighbors, who may or may not have seen the same cable program. The social distance from the television viewing experience encouraged selective discussion of content and placed a particular spin on an issue. This process brought the earlier level of participation into the realm of a shared discourse, where issues entered the Woodglen public forum and were subject to its tacit rules. As the fluid pool of residents showed its views, which were similar and varied in different respects, Woodglen's rules exerted their influence. A conventional Woodglen wisdom, not without its internal contradictions, appeared to emerge. With Lila's and Emily's experiences as examples, these two levels of public participation can be explained.

Lila and Emily both told me that they perceived the political process as being completely inseparable from television. They maintained their own face-to-face involvement in political affairs through their work with the League of Women Voters and as volunteers for the county elections office, assisting with voter registration drives and working in polling places. They said they had observed that the generation of knowledge about and interest in politics as a whole—for themselves and for others—proceeded in varying degrees from television. This was a condition they identified as an outgrowth of the medium's expanding technological landscape—with emphasis on the proliferation of specialized news channels. Emily and Lila found these many new viewing choices an exciting indicator of a distinctly new period of public interest in politics.

By plugging into journalists' roundtable discussions, C-SPAN's event coverage, and any of a growing number of cable talk and call-in shows, the two believed, they could more directly "participate" in national politics than when they received their information through formats that reflected more distance—through space and time—from their point of origin. "Now, I am more directly in touch with what is going on, and, because I understand it better, I am more able to do something about it, through voting, through legislative contact, or through talking with other voters," Lila said. These two women said they encountered the barrier of distance from events through conventional network styles and packaging that manipulated the news product, including such techniques as short news stories that distort issues and an emphasis on the professional journalistic value of detachment. In contrast, they believed the new formats encouraged public dialogue and enlivened the democratic process by inviting viewers closer to reality.

"The call-in shows are a barometer of a sort, and they offer another perspective from mine," Lila said. Emily added: "I am very interested in what's out there, and the kinds of questions people will ask." Both told me that talk and call-in formats, which encourage expression of public opinion, helped to make a difference in citizen involvement in the 1992 national elections. Lila observed: "In our county, we had a record voter registration turnout. I think television such as this has a lot to do with it. You cannot avoid the talk shows."

Contrary to a traditional view about television, that it cannot provide the depth that constituents can get from print media, Lila and Emily see different results from the proliferation of cable channels. Emily reflected on some of the traditional pressures about how time should be spent and expressed some doubts about whether television represents a serious enough leisure pursuit: "The time you spend watching politics on television, I feel guilty about that. There is so much reading to do. With television, though, there's so much more detail you can get, and you find a cross section of public opinion."

"I don't feel guilty about it," Lila countered. "I realize how much more an informed voter I have become. I think it's a great indoor sport, too. Anticipating the presidential debate is a little like looking forward to the big game." She enjoyed television in a context of leisure but felt the act was a serious one.

Lila observed that new technologies have influenced involvement in local politics in a way that is similar to influences in the national arena and in a way that is different. On a day when we met, Lila's League of Women Voters chapter was to hold a forum for local candidates for election. Admitting a twinge of regret, Lila told me she would not attend, because she intended to stay home and watch George Bush on *Larry King Live*. Influencing her decision was the league's arrangement to narrowcast the forum on the local cable system's public-access channel the following evening. She commented: "I suppose it works both ways. People stay home from the event because they know they can see it on television, and more people wind up seeing it *because* it's on television. Cable has changed public participation."

The reconstitution Lila perceived within League of Women Voters forum participation was closer to what was happening on the national level than on the state. She and Emily said cable was missing an opportunity for constituents to participate in state politics and they, as newcomers to that political scene, were ignorant of it as a result. Benedict Anderson has theorized that television has helped to construct an imagined community of Americans who will never meet one another but who share a notion of what it means to be part of a nation.[11] In their obsession with U.S. domestic and foreign news matters, the Woodglen cable "junkies" similarly impressed me as belonging more to such a national community than to a state or local one. Many of them had lived in the state for much or all of their lives, although others had spent their lives—career years—elsewhere. Most had cut or relaxed long-term ties with people they had known intimately for most of their lives; they moved away from neighbors and co-workers. For women, who are socialized to maintain permanence in relationships, such cutting of ties may have been especially difficult. Then, to find themselves permanently cloistered in a setting in which civility, rather than intimacy, was encouraged, their interest in the national "community" was encouraged at the expense of membership in a state or local community, where relatively few of them shared a history.

Emily and Lila said they felt their personal interaction with television was linked with their considerations of content among contemporaries at Woodglen. They shared information they collected from television, compared notes with neighbors, and learned about content they missed. The news and perspectives they assembled from their time with television provided fodder for communal table conversation among friends and served as icebreakers in less intimate settings.

A Tube of Activism?

For Barbara, political communication monopolized everyday life, whether through fact gathering, using media as tools; perspective sharing in her community; or acting personally, using her pen or checkbook or doing volunteer work. She and many of her neighbors told me that they wrote occasional letters to elected Washington officials, pushing their positions on particular bills. Some of them belonged to lobbying groups—Barbara held membership in several environmentalist organizations and sent them regular checks. They were in her will. The manifestation of Barbara's own brand of politics was inextricably linked to her digestion of media. Its complexity contradicted the popular view that television oversimplifies issues and leaves people little room to think for themselves. Barbara's own facile use of media capitalized on the comprehensive and diverse content that she believed cable could offer in her effort to understand politics.

One fall day in 1992, I met Barbara in her flower garden, several months after she had told me of collecting signatures for Ross Perot's entry onto the state ballot for president. That day, she told me: "I had a great deal of aspiration for Perot, although I have been disappointed by him." Perot's quitting and then reentering the race disappointed her, as did what she observed as his progressively unrealistic position on dealing with Washington "gridlock." By the autumn, Barbara had thrown her support behind Bill Clinton and Al Gore, the Democratic candidates, whom she found to be humanitarian people with decent reform proposals. (She had been a Gore enthusiast since well before the primary season; she liked his environmentalist views and had read his best-selling book.) Even as she prepared to cast her vote for Clinton and Gore, Barbara told me that larger issues than the election loomed that she was not so sure the pair could solve: "I'm terribly concerned about our children—*your* children. I don't want them to wake up one day with a situation—the deficit—that they cannot control. You've just had it."

I heard many such statements during my two years of visits to Woodglen from both men and women there. Barbara's level of concern with the democratic process, as heightened by her monitoring of proceedings and developments with television, may have been extreme even for the community. However, her statement about concern over the national debt resonated with a theme I encountered many times there: the feeling that there was some control to be had over Washington, that involvement by residents could be constructive. People reported to me that they attempted their involvement through traditional means, such as voting and financial support for causes, but just as commonly, this involvement took place on the mundane level of everyday talk in the public spaces of Woodglen. Residents there recognized that their television viewing, with its connection to their focus on living in the present, the search for quality, and a serious attitude toward leisure, bore the stamp of productivity in the public sphere.

Conversely, something could be known about these elders because of the television viewing they did not do. Most of them, for example, avoided low-prestige genres (often linked with female viewing) such as television movies and situation comedies, where the focus tended to be on the private sphere. As illustrated by the comments of Midge, the single admitted soap opera fan, any talk I did hear at Woodglen that was favorable toward entertainment television focused almost entirely on concerns of the traditionally male public sphere—soap operas as a means to educate the public about health, Murphy Brown seen as a public leader (or symbolic enemy), and so on. I found this odd, because these television texts so often focus on the realm of the private, traditionally female, sphere.[12]

The Woodglen community, although populated primarily by women, was dominated by an upper-class male perspective that was informed generally by the class position that its residents occupied and specifically, I believe, by the residents' places in the professional and academic world. This perspective pressed residents to value the foregrounding of public matters and the suppression of private ones. In the most obvious example, conversation between residents in the public spaces of Woodglen focused on politics and cultural events but not on one's health and on one's children, as we have come to expect older people's talk to be constituted.

The elders at Woodglen took a serious pleasure in their political involvement through their manipulation of television. However, activity is not the same as activism. I believe the public participation that occurred there was circumscribed within the framework of liberal democratic politics as it could be understood from virtually all the television texts available for their consumption. This framework inherently endorsed the dominance of the American nation-state in public life. This somewhat contradictory experience among the Woodglenners—using television to participate in activism that was inherently limited by the terms of the available political discourse—makes them no different from other Americans generally who derive their public knowledge from mainstream media, whether popular or middle-brow outlets. While the outcomes of their community politics may be already inscribed within a familiar range of possibilities, I am heartened by their zeal for participation.

Woodglen Logic and News Show Hierarchy

Not every Woodglenner, of course, depended on television to ensure successful participation. Frank and Charlotte, whose television use was minimal, were the Woodglen ideal. In their 80s, they enjoyed reasonably fair health and an active lifestyle. They had lived long, fulfilling, highly cultured lives, and their children held important jobs in major North American cities. Both Phi Beta Kappas and Harvard and Wellesley graduates respectively, Frank and Charlotte remained active in academic life, where Frank had spent his career years as a philosophy professor. He continued to publish his work, and both associated socially with friends from the local university. Frank continued to drive, and the couple attended cultural events frequently on the nearby campus.

Their upstairs apartment in Woodglen's main building was filled with imposing antiques and a massive book collection. Instead of a dining room, they had made a space for Frank's study. They had let go of most of their entertainment trappings. Hidden away in the bedroom, in a corner a few feet from the huge, postered bed, was their tiny television set. For little more than two hours per week, they devotedly sat on straight-back chairs and watched PBS's *Mystery!* and *Masterpiece Theatre*, cordially shouting at each other, "What did he say?" because they were hard of hearing and unable to understand many bits of dialogue. They largely despised television but were grateful to have front-row seats to "the theater" at a time in their lives when they no longer lived in a big city where they could attend many plays. Very rarely, they watched a news program or documentary on either public television or a cable channel.

In no way did Charlotte and Frank live a media-poor life. They spent a couple of hours a day wading through the *Washington Post* and sometimes the *Chicago Tribune*. They bought the *New York Times* on Sunday, although they considered it a burden to pore through. And they received the *New Yorker*. They were highly committed liberal Washington observers and considered television mostly a vulgar conveyance of irrelevant material. The only news coverage they faithfully tuned in for was presidential and vice presidential debates and party conventions.

It was in these instances, Frank noted, that television had always shone.

Charlotte added: "In the vice presidential debate this year [1992], Mr. Bentsen's quick reaction and Mr. Quayle's unfortunate facial expressions just wouldn't have come out on the printed page."

"This is entirely apart from any feelings of suspicion we might have of television giving one-sided or overly simplified representation," Frank said.

In this atmosphere—where the dominant ideal of the elite flourished—television use was conducted at Woodglen. A good many residents did not watch any more television than Frank and Charlotte did, and a few did not even own a set. However, even among the residents who had large console sets positioned in their living rooms and who watched them frequently, viewing habits bore the influences of an upper-class bias against the medium. Many residents told me, almost sheepishly, that they were not interested in the high-culture entertainment offerings of PBS, as many of their neighbors were, but they liked to watch the news.

It is understandable, given Woodglenners' background, that news coverage that some viewers find laborious and tedious carried special appeal in the community. In order to understand why, it is helpful to appreciate some things about the identity of this group. Most obviously, their membership in the well-educated upper middle class afforded them a range of twentieth-century Western history that provided a cultural passport for understanding complex news events. Many of them told me that their storehouse of knowledge caused them to dislike having events oversimplified or explained out of context by journalists whom they perceived as telling stories according to an audience formula geared toward the lowest common denominator, a formula they associated with popular news shows. They preferred the brooding financial analysis of PBS's *Wall Street Week* or the sprawling technical discussions of C-SPAN. As Frank explained to me:

> On the evening news, very often, they will flash a quick graphic on the screen telling you three or four points about a subject, such as the health care reform package. And the story, which must be quick in order to fit the half hour with all the all-important advertising, does not stray very far from this neat graphic. Well, that hardly tells me anything about the issue, does it? I want to know the *history* of the thing, its implications.

I found the Woodglenners' news and public affairs program preferences to be positioned along three axes: a shared macro (versus micro) view of events, a preference for investment versus consumption, and a liking of the logical mode of television versus the medium's oral mode.

Macro versus Micro View of Events

Residents told me they wanted to grasp the big picture. They wanted to know the impact of events and how the parts of issues worked interdependently. For the Woodglen viewer, single events were always tied to larger matters. They preferred news shows that were geared toward deeper questions. For example, viewers often volunteered to me that they connected the *events* of the Clarence Thomas Supreme Court confirmation hearings and the William Kennedy Smith rape trial, telecast by CNN in 1992 and 1993 and widely watched in the community, to broader social and political *issues*. Likewise, many watched the Lorena Bobbitt trial telecast in 1994. "I found the trial to be of enormous significance," lifelong homemaker and civic volunteer Geraldine told me of the Bobbitt trial, which she had watched on a ten-inch screen in her dining room. Her husband, Sam, a retired shop owner, added: "This case is very relevant, given the issues about abuse of women in marriage." The couple followed continuing coverage on CNN and in news magazines and told me they looked forward to reading the *New Yorker*'s commentary on the subject. Like their neighbors, Geraldine and Sam mentioned the story to me only in terms of its social significance, never in terms of, say, the racy, man-bites-dog quality of a news story about a wife cutting off her husband's penis. My experience was the same with viewership having to do with the O.J. Simpson matter: residents exclusively expressed their curiosity in terms of the larger questions, such as race relations and the inadequacy of laws designed to protect battered women.

Even though programs such as tabloid shows may well have connected such events to issues, Woodglenners shunned such outlets because they found them lacking in legitimacy. They liked seeing the "expert" sources on CNN and similar milieus: national women's group leaders and trial law academics who could help frame the stories, for instance. Tabloid programs were more likely to rely on ordinary people such as story subjects' co-workers or social acquaintances to help construct their narratives, framing the issues through inflammatory commentary. Woodglen residents' recognition of officials and other experts as legitimate sources for the framing of events as issues suggests that their decision to seek such issue-based content was ideologically based. They trusted that the most credible framing could take place with the participation of people who were highly placed in powerful institutions—reputable, responsible, conservatively dressed people who seemed personally detached and, perhaps like themselves, carried elite academic credentials.

Media critics have tended to suggest that journalism constructs the frame through which audiences understand issues, but my observations of people at Woodglen suggested that these viewers at least exercised a certain amount of creativity in this process through their combing of news offerings for various pieces on subjects of interest to them. Their impulse to seek out news analyses stimulated—and was stimulated by—their tendency to see events as connected to large webs of issues. For example, residents told me they avoided quick daily reports on the status of conflict in Somalia or Bosnia but searched for discussions with panels of experts who hashed out implications for U.S. foreign policy in light of such events. Or, instead of concentrating on a network White House reporter's stand-up announcement of the day's Whitewater scandal news, they preferred to watch Ted Koppel plumb the depths of "the issue" and then tune in to a cable channel's analysis of the investigation as it was informed by the costs and benefits of the Iran-Contra special investigation.

Residents told me they liked programs and journalists that projected a global perspective. When pushed, they asserted a primary concern with U.S. interests but said they wanted to know more about world affairs than they could glean from network news reports, which they called overly "domestic." One woman said she liked listening to translations of Moscow newscasts on C-SPAN, because they helped her "fill in the picture from Washington." Another resident, Don, said he liked Peter Jennings's ABC *World News Tonight* over its competitors because the host, a Canadian, showed "complete objectivity." Individuals apparently felt they could find a more macro news perspective on shows that looked past U.S. borders somehow for information.

Investment versus Consumption

Woodglenners chose programs that they said represented a high rate of return for the time they invested. At the high end of the axis were representations of high-culture performances and such shows as PBS's *MacNeil-Lehrer News Hour*. At the low end were programs that residents judged as requiring mere consumption of time and of worthless content—mere entertainment. Such programs would include tabloid television and most fictional television outside the productions of PBS.

Among the highest investments residents felt they could make were documentaries, especially those on public television. The content of documentaries often formulated material for discussion in the public spaces of Woodglen, especially the dining hall, where people felt it was important to have something both serious and topical to say. What residents told me about their experiences with these "powerful" and "revealing" programs resonated with media scholar Bill Nichols's view of documentary—that it is connected to "the discourses of sobriety—economics, politics, foreign policy, education, religion, welfare"—systems that have the power to effect change in the world.[13] Residents told me many times that they liked both documentaries and public affairs shows because they liked to "talk about the issues of the day," and they considered the framing of historical events a set of issues in itself. Nichols has identified *epistephilia*, or desire for knowledge, as the basic component of documentary viewership, and, in the case of Woodglen, this condition can be extended to other nonfiction genres. Nichols says that documentary:

> posits an organizing agency that possesses information and knowledge, a text that conveys it, and a subject who will gain it. He-who-knows (the agency is usually masculine) will share that knowledge with those who wish to know; they, too, can take the place of the subject-who-knows. Knowledge, as much or more than the imaginary identification between viewer and fictional character, promises the viewer a sense of plenitude or self-sufficiency.[14]

The pleasure that Woodglen viewers said they took in "knowing" was fed through an elaborate information network that they had woven for themselves from television's various products and from other media. Many Woodglenners told me they were simply used to learning. "Learning from TV makes me feel powerful, in a way," Lila commented in a focus group meeting. "I don't have to lie down and die just because I'm a so-called senior citizen." Retired schoolteacher Verna, in her mid-80s, added: "Use it or lose it. That's the kind of thinking that brought us to this place."

I noted a special poignancy about the use of television by such people as Lila and Verna, who compulsively looked to beat the odds against aging. Their epistephilic urges were a means of staving off inevitable death, or perhaps properly investing one's time until it came. "If you're not living, you're dying," one woman, Fostine, once told me. Barbara later put it differently: "There is always somebody dying here. When I look out and see what is going to happen to my carcass, well, it's depressing. I turn to the world outside through the information of television as a means of avoiding it."

For all the residents I interviewed, high-quality nonfiction television seemed just to make "good sense." Most of them came of age prior to the Great Depression and were rearing children during this time of hardship. Although they suffered less than blue-collar families, they either continued or adopted frugal spending habits and never threw themselves headlong into the consumer culture that boomed after World War II. Even among residents who had been involved in the professional world, material possessions had been acquired to last.

They looked upon television in a similar fashion. Praising its capacity to make them smart and condemning its capacity to waste their time, they were quick to divide programs into "good" and "bad" categories. The "commercial" channels—minus CNN, because of its tendency to play up issues and "downplay journalists' personalities"—generally meant "bad." The more that television afforded them the opportunity to "invest" time in content that did not bring them so much personal gain but a fulfillment of their sense of duty to know the contemporary world, the better they liked it. Frances, a retired library science academician from a different state, told me she watched very few programs and selected them carefully. Her favorites were the PBS news programs *The MacNeil-Lehrer News Hour* and *Washington Week in Review*. She said both were worth her time: "They make an attempt to bring together more than one viewpoint. They don't bring all viewpoints in—we can't demand that, for heaven's sake. But it's not like the network man-on-the-street. These people are carefully chosen for their positions and backgrounds. It helps me to work through complicated issues that are glossed over in other places." Ironically, what Woodglen residents perceived as an investment of their time can be argued to have been consumption. As Roger Silverstone puts it, we can see goods as symbols, symbols as goods. Once symbolic goods enter a system of commodification, such as television, they take on an exchange value.[15] For Woodglenners to use television for their advantage—as an investment—was in itself, of course, an act of consumption. They were interested in consuming those texts that had value in their social world, a criterion that seems to be shared by most people in choosing what to watch on television.

At Woodglen, residents regard investment as an active use of television and consumption as a passive use. This distinction blurs when we consider that these viewers deliberately seek and use particular contents for their local use value. Although I observed a chain of examples in which residents sought out information from television that they could then employ as fodder for mealtime talk at Woodglen, I am not suggesting that they had no other reason for wanting to view such material. Collecting and sharing information, however, constituted a prudent investment of their time even as it constituted a certain consumption of the television product. Beyond this collecting and sharing, another level of consumption was at work: residents often structured their days to accommodate certain programs, such as *The MacNeil-Lehrer News Hour* or even *Jeopardy*, the game show for smarties, which is ritually viewed by older people I have met from various social backgrounds. As with soap operas for homemakers and college students, these favorite information programs influenced when and how Woodglenners arranged to be at home to have their television needs satisfied.

The Woodglen elders tended to refer to "worthy" or "wasteful" television, often valuing "public" or "cable" over "commercial" television (meaning mostly network television). Because it was devoted entirely to news, CNN was not linked with the "commercial" channels, although residents were aware of its advertising content and disliked it.

The residents' easy distinction between "good" public-focused television and "bad" commercial television muddies when one considers the limits of the American model of television. PBS may not be known for commercial interruption, but it has its corporate sponsors; C-SPAN has no commercials, but it operates within an ideology. Daniel Hallin has argued that television assigns ideological meaning to political events. Tracing television's ideology to the liberalism of the Progressive era, Hallin notes:

> It is reformist, and stands ready to expose violations of ideals of fair play, equal opportunity, and prosperity. It is also populist in the sense that it is suspicious of power and those who seek it. . . . But reformism and populism coexist with a strong belief in order, consensus, moderation, leadership,

> and the basic soundness of American institutions and benevolence of American world leadership.[16]

These principles fit the worldview of Woodglen residents, most of whom revealed reformist and populist sympathies in my observations of them. For example, Emily watched the Clarence Thomas hearings on CNN with great interest. She, like many of her neighbors, told me that she was concerned that Thomas, whom she believed guilty of sexual harassment, apparently was being elevated to a position of great power after having behaved both illegally and "shamefully." Verna, an ardent Republican, watched the hearings with "disgust," believing that Anita Hill had been put up to her accusations by the Democrats. Verna believed party politics was jeopardizing the institution of the high court and should be preempted.

Emily and Verna, while disagreeing on an issue, saw, through television, the potential for individuals to prevail against wrongdoers from within the establishment. Both believed that television revealed the relevance of the Senate hearings. "CNN showed how the American system works," Verna noted. Regardless of who emerged as the "winner," both were satisfied that the CNN cameras reflected the same America they knew, in which institutions—the Senate and the Supreme Court, in this case—may be damaged by wrongdoers but correctly exist in society. As the Glasgow Media Group has claimed, the mass media as a whole tend to reinforce powerful interests: "Information is both controlled and routinely organised to fit within a set of assumptions about how the world works and how it ought to work. The media relay the ideology appropriate to a population which is relatively quiescent, and actually promote that quiescence by limiting access to alternatives."[17] Such influences on the elite programs favored at Woodglen serve ultimately to reinforce a traditional worldview even as they appear to open up a choice of views.

The homogeneity of the Woodglen group presents an empirical case in which viewers' subjectivities keep them from developing perspectives that are very different from those that seemed to be encouraged by elite texts. Woodglenners, for example, watched *Wall Street Week* because they did not fundamentally question the correctness of a market economy—indeed, they participated in it—and they watched C-SPAN because they fundamentally believed in a centrist government. Additionally, the same types of programs were popular throughout the community. This textual similarity and shared consumption prevented people from sharply contesting the meanings they got from programs. People had no need to form understandings that differed notably from the ones inscribed in the texts.

John Fiske's argument that texts bear the potential for audiences to decode them more or less democratically through the exercise of their own subjectivities is useful in considering the pleasure that a predominantly female audience derived from male-dominated texts.[18] For instance, in viewing C-SPAN, the women at Woodglen were watching a field of mostly male participants engage in a traditionally male realm of national politics. To be sure, women at Woodglen may have accepted meanings as they were constructed by the producers of complex news texts, but they were perceiving themselves as powerful in selecting from an array of meanings that were available. For many of them, since their husbands' deaths, their own retirement, and changes in television's technology (e.g., remote control and cable expansion), they felt they were making their own "investments" for the first time. Several women echoed, in some fashion, the comments of Barbara:

> Even though I was a professional, it was my husband who tended to espouse the major political views in the family, and he made most of the decisions. He certainly made the decisions about what was watched on television, and it was sports most of the time. I had no interest in it. I had no knowledge of the political environment, really. Now, I am alone, and I have discovered television, particularly C-SPAN. And I have my remote control. It is raw power.

Barbara made the comment about her remote control only half jokingly. She said she felt as if television took her places at a time in her life when she had had to give up most physical travel, which she found uncomfortable; this is a theme I heard echoed by many men and women I met in the community and elsewhere. Barbara and several other women told me television made them feel powerful in ways they had never imagined; they got pleasure through their participation in such channels as C-SPAN, with its traditionally male-oriented texts. Such a view substantiated Fiske's position that people play with television's meanings to suit their own subjectivities. While Barbara, Emily, and Lila, for example, recognized U.S. politics as a realm in which their husbands had been involved, they attempted to use texts from this realm in order to make themselves more powerful—to locate themselves within the male arena, challenging the notion that politics is a site that is closed to women.

Formal Logic versus Oral Logic

John Fiske and John Hartley have argued that television, with its logic of pictures and rhetoric, presents itself more in an oral mode than in a formal, literate one, which is based on, among other things, permanence of presentation, abstract presentation, and individual reception. Storytellers work in the oral mode, and written documents such as books exercise the formal, literate logic. Fiske and Hartley persuasively suggest that television's "meanings are arrived at through the devices of spoken discourse fused with visual images, rather than through the structures of formal logic" but that the medium's "final form results from the active tension between" the oral and literate modes.[19] Television, while using the spoken words and pictures of the oral mode, nevertheless transmits a cultural logic through the literate mode in the sense that it is transmitted from "highly literate specialists, from newsreaders to scientific and artistic experts."[20] Audiences encounter and negotiate this logic through their own experience with culture.

The oral logic that Fiske and Hartley find in televised programs, especially news, is less prominent in the programs that are valued in the Woodglen community. C-SPAN, for example, although certainly filled with speech and pictures, favors a formal logic. Its messages are presented linearly and logically, with events being coded on screen so as to furnish a "record" for the viewer: congressional votes are tallied on screen, for instance, and the viewer has access to a video scorebook in order to comprehend how such events have occurred. This tedious construction of events occurs sequentially. No story superstructure exists, but the House actions as presented through C-SPAN have a narrative quality that is absent from network news shows. The House actions are presented in their official entirety, not in abbreviated, dramatized form. For example, on March 17, 1994, the House of Representatives debated a bill that would have outlawed obstruction of people entering abortion clinics in the United States. C-SPAN's coverage presented the debate "from gavel to gavel" along with a tally of the resultant vote. Network news coverage, however, included a video clip of abortion clinic protesters, which dramatized the event. Woodglen viewers valued C-SPAN because they believed that it presented events in accurate order with a consistency that allowed them to apprehend "truth" for themselves.

The Woodglenners' preference for a literate logic of television extended to their impression that they could locate a totality of texts, through their own industrious efforts, that allowed them to understand complex events. By turning to cable news channels and "serious" news programs, they were not limited to quick, ephemeral accounts. For example, at a dinner meeting one night in the fall of 1992, Daphne, Rachel, Gloria, and Barbara talked to me about the excitement they felt in seeing other nations' parliamentary proceedings on C-SPAN as well as telecasts of indigenous news programs from Russia and a variety of Asian countries. They mentioned proceedings from Britain, Canada, and Australia. "I'm interested in seeing them discuss the problems there are before them, not just learning the American journalists' versions of them," Rachel noted.

Woodglenners described for me a combination of programming preferences, including open-ended and complex discussion, a perception of minimal processing of "the facts," an opportunity to "witness events firsthand" through a camera that was perceived as impartial, and the presence of authoritative experts who give dimension to issues. In order to fill these needs, residents appeared to seek out programs that featured a formal logic, programs that told a story as completely and sequentially as possible and appeared to invite their reflections.

The Woodglen Critical Logic

Residents of the community seemed to have no trouble distinguishing a program's suitability for their viewing based on their standards of quality. Acceptable, at Woodglen, did not mean pedestrian. For example, residents overwhelmingly expressed to me their loyalty to such programs as *The MacNeil-Lehrer News Hour* and *Washington Week in Review*, but only about a fourth of them told me they watched *60 Minutes* or a network evening news show. The tabloid news magazines may attract many older American viewers, but they did not draw them at Woodglen. I came to recognize two ingredients that especially influenced the appeal of news programs among residents: lack of commercial interruption and the appearance of high journalistic standards.

Commercials and Television

Very few Woodglenners told me they tolerated programs with commercials; even those who enjoyed CNN said they often used their remote control devices to plug into noncommercial channels when the ads came on. "The commercials are so loud," Frank told me. "And they don't make any sense at all. They're great annoyances which we simply do not care to abide."

The appearance of commercials made Woodglenners sense they were wasting their time, and such content was the chief reason people named when they told me they avoided television entirely. A retired schoolteacher, Verna, told me that she often watched a bit of NBC's *Today Show* in the mornings but did not dare keep the set on long: "I turn it off. There's just too much money involved. It's all hype for money." One retired businessman, Don, was a devoted television watcher but found the commercials peskily superficial: "These beer commercials with the scantily clad women—how stupid can you get? Like that Budweiser commercial about the guy in the limo—what is the substance of that?" Another resident told me he found the commercial content "shameless." Some residents expressed fears of a third-person effect that advertising might have; most feared that, although they were not gullible with respect to commercial claims, many older viewers were.

Several residents told me they circumvented the problem of commercials through time shifting. As Lila put it: "Some of us record the program and zip right through."

Journalistic Legitimacy

Completeness, fairness, and balance are hallmarks of American journalism that Woodglenners found essential elements in high-quality shows. The more information they could get from the most well-informed subjects, the better they found the program to be. Woodglenners' television literacy allowed them to recognize tabloid shows as "trashy." As people described this genre to me, they tended to mention the lack of "serious" journalists, sensationalism, gaudy music accompanying stories, and a practice of "re-creating" news events. "Most of us steer clear of these programs," Verna told me.

Generally, people told me that they found conventional news programs, such as CBS's long-running news magazine *60 Minutes*, more respectable than tabloid television but still lacking the depth and professional detachment that they expected from journalists. In such shows, residents said they recognized an attempt at objectivity, coverage of a broad range of serious subjects, "star-quality" professional journalists (whom they knew to be highly paid), and high-quality production work, such as good informational graphics. Not all Woodglenners like these popular news shows, however; some sharply criticized their bias and "amateurism." Such criticisms included chides against Dan Rather's odd "boorishness" and incidents of unprofessionalism (such as his walking off the set in Miami several years ago after *The CBS Evening News* was delayed by a sports event); the unprofessional treatment of his colleagues by *The Today Show*'s Bryant Gumbel; and the silly crotchetiness of *60 Minutes*'s Andy Rooney.[21]

Among television news personalities, residents are drawn instead to C-SPAN's founder and sometime host Brian Lamb, NBC's Jane Pauley, CNN's Larry King, and ABC's Ted Koppel. Lamb was revered for his dogged neutrality, even his blandness. Pauley they found simply to be a no-nonsense journalist with a "professional style," not "silly" like the morning news show anchors who had come after her. (They had been especially repulsed by the beautiful Deborah Norville.) King—so different from the low-key Lamb—was oddly popular at Woodglen. With his notoriously soft interview style—the opposite of Koppel's persistent adversary role—King never put guests on the spot and never achieved a reputation for research. Residents nevertheless found King's demeanor "refreshingly polite," even "classy," because he unfailingly treated guests with dignity. Woodglenners embraced King as well as Koppel because, although they had very different styles, neither seemed distasteful.

One explanation for Woodglenners' acceptance of both Koppel and King, and their rejection of such anchors as Rather and Gumbel, is rooted in the personality ideal of television news anchors. Daniel Hallin has observed that these individuals logically form "a persona that combines authority with likability."[22] Hallin's example of Walter Cronkite is the ideal type of the network anchor persona. Like Cronkite, Koppel and King were perceived at Woodglen as both authoritative and likable, whereas Rather and Gumbel were perceived as amateurish and "smart-alecky."

Among the people who watched a network news show in the neighborhood, most tended to say they needed to augment their information with specialized coverage from news talk shows and the like. "We need this information to talk about the issues," Lila said. "We get our information from television, and we feed on each other."

Many Woodglenners said they had long watched *60 Minutes*, but several of them had stopped in recent years. Residents generally said they disliked the show's theatrical journalism, its lack of evenhandedness in investigating news subjects, and too much personalizing of news by its celebrity reporters. They disliked what they perceived as a lack of depth and the advertising that surrounded and interrupted it. As a woman in her 70s, Gloria, told me: "I used to watch it, but I stopped years ago. You know what's going to happen by the time you see what they're up to. If they're after someone, they're going to do a hatchet job. If they're doing a profile, it'll be a puff piece. Count on it. And Andy Rooney, quite frankly, has worn thin. I thought Shana Alexander and Jim Kilpatrick at least had something to say." What may have made *60 Minutes* appealing to many older viewers—its routinized framework, predictable conclusions, and what Richard Campbell has identified as its comfortable middle ground—condemned it for Woodglenners.[23] They felt patronized by the program's tendency to cleave issues into simple pros and cons and especially disliked the reporters' flashiness. As Verna noted about Mike Wallace: "The minute you see him, you know what it's going to be. He's going to get them, whoever they are. He's kind of a hot dog, Mike Wallace is. He's always in the foreground, always so smart. It's a pointed view of the subject one gets from *60 Minutes*." Among those who remained loyal viewers of *60 Minutes* at Woodglen, they tended to say that the program was inferior to others that they preferred, such as the *MacNeil-Lehrer* show. They continued to watch *60 Minutes*, however, because of its biographies of famous people, especially older celebrities.

Television News and Other Talk

The Woodglen communal logic with respect to television translated into a preference for programs that amplified concerns with the public and suppressed concerns with the personal. The news genre primarily highlights a concern with the public, although its subgenres and the channels on which they appear inject various degrees of the personal into their formats. For example, C-SPAN's coverage of congressional proceedings is arguably the least personal news coverage on television, and the most public, because there is no narration, no address of the audience (outside the call-in show), and no narrative construction. Camera direction and minimalist labeling of events superimposed onto videotape form the crux of its coverage. It appears to be a "presentation" rather than a "representation" of a public event, and Woodglenners warmed to it because of this. On the other hand, a chatty morning network news show, shot on a set meant to recall a middle-class American home, includes personal talk between hosts who seem to represent television's ideal of the American married couple, who relate news from the public realm but privilege the telling of stories from the realm of the personal: celebrity interviews and consumer how-to stories.

Sonia Livingstone and Peter Lunt have noted that public television traditionally has concerned itself with the viewer as citizen, while private, commercial television has concerned itself with the viewer as consumer. The authors conclude that these terms frame public television as elitist, with viewers detached from its discourses, while commercial television's viewers are involved with it. Livingstone and Lunt have identified a new area of television, representing a constellation of genres, that they consider to constitute "participatory" television. They have located this genre in the "interstices" of the public and commercial models and believe it to have a potential emancipatory role, encouraging democratic access to public discussion.[24] The authors primarily are concerned with the new audience discussion programs of television, including *Oprah!*, in which "ordinary citizens" join "experts" in a studio to debate issues through the mediation of a host. To these programs I would add audience call-in shows, some of which have "experts" in the studio to field questions and others of which have only a host to hear comments and offer replies. Shows with studio audiences and callers are similar to call-in shows without audiences in some ways. For example, both have "experts" in the studio (or otherwise on camera), both are mediated by a host, and both may offer access to long-distance callers. However, crucial differences exist. Unlike shows that feature studio audiences, which are often taped, call-in programs such as *Larry King Live* are telecast live, encouraging viewers' impression that they are observing reality in a more or less unmediated manner and that they are able to, in fact, participate in the event. The absence of a studio audience emphasizes the national or global nature of the call-in audience. More than one viewer at Woodglen told me that they truly felt that such programs made them feel a part of "the global village."[25]

Livingstone and Lunt suggest—and Woodglenners seemed to concur—that participatory shows "channel public opinion up to the decision makers and . . . circulate it among the public . . . [providing] a social space for the public to debate issues of general concern." The elders at Woodglen tended to tell me that they favored programs featuring the greatest amount of variety among "ordinary" guests with the least amount of interference from hosts and experts. It is in this most participatory mode that they said they perceived democracy to be most effectively at work. At the same time, these elders tended to watch mostly call-in shows that were oriented toward public issues. Call-in segments that express the personal, rather than suppress it, such as those of the two home shopping channels that aired on the Woodglen cable service, were shunned in the community. While viewers said they valued the voices of "ordinary Americans," they favored shows that featured the guests and journalists whom they found to be the most credible and authoritative (Jim Lehrer more than Peter Jennings, Peter Jennings more than Dan Rather).

As in the case of soap operas, I am suggesting here that a class- and gender-related bias contributes to the Woodglen community's preference for more public-focused discussion programs rather than the more personal *Oprah!* types of programs. Livingstone and Lunt have suggested that the audience participation shows they studied represent a women's genre because they often concern personal relationships, gossip, and storytelling, and they are scheduled during the part of the day that is devoted to housewives.[26] The talk and call-in shows that were popular at Woodglen, however, may in turn constitute a men's genre, because they focus on national public issues and debate in place of gossip and storytelling, and they traditionally are scheduled in the evening, a part of the day that traditionally has catered to men. It is significant that a community of primarily women at Woodglen selected a genre of programming—the news-oriented talk show—that concerns the public sphere and a heretofore male perspective. As with the case of soap operas, this is a case of class identification superseding both gender and age identification. High levels of education and economic capital have turned these women away from traditionally female genres. Some of the women told me that they had no interest in talk shows that focused on interpersonal relationships, and many told me that they found much of the daytime television content outlandish and offensive.

Soap operas, a genre traditionally associated with women, are spurned at Woodglen like another traditionally female activity, gossip. Although I did hear gossip from several women while I was in the community, even from people who frowned on other women for gossiping, the occasions were rare, at least for me in my capacity as an outsider and observer. The rejection of these conventional outlets for women was made doubly obvious by the women's favoring of more traditionally masculine outlets, such as talk of politics and college sports and general knowledge about one's field or the world. I largely attribute these masculine-sphere preferences to these women's class, education, and professional backgrounds. Most of them had never watched soap operas. Although many had been homemakers in their younger years, they had also participated in activities outside the home. None seemed to identify with the homemaker role in retirement.

Television with Purpose

I remain amazed at the level of purposefulness with which Woodglenners approached television, partly because of my own lifelong habit of taking the medium's place in my life largely for granted and because of my expectation that most people do just about the same. These elders skillfully integrated the medium into their everyday lives to the extent that they were able to rely on it to help them fit the demands of the discursive template to which they felt bound. A man watches PBS's concert with the world's most famous tenors not because he particularly enjoys it but because he knows his dinner companions the next day will consider it worthy of discussion. A woman switches on *Larry King Live* in the evening because her neighbor mentions that she has read somewhere that attorney general nominee Zoë Baird will take phone calls from the public. In retirement, and perhaps after the death of a spouse, these elders exercised a new means toward what they had come to consider worthwhile pursuits, thinking about public-sphere action as presented through the fluid pool of television texts that were available to them. Their actions were continually reconfirmed for them through their community.

People living in communities such as Woodglen may use television in their construction of what Haim Hazan has identified as an "alternative reality," which is not anchored in the real linear movement of time.[27] The residents of Woodglen did not consider their lives in terms of movement from the past, which was behind them, into the future, whose prospects for them were dim. Instead, many used nonfiction television to participate in a community discourse that unfailingly privileged the present. They believed they "witnessed" reality every day with constant "live" access to events. Through television, they felt unbound by geographical restrictions and more a part of the national community. This involvement in television helped to constantly rearticulate Woodglenners' understanding of time and space in terms that helped them to feel vital and comfortable.

Four

Preaching to the Unseen Choir

Johnny Dixon joyously waves the observer into a tiny black foam-covered room where he is walking about near a video camera. He is 70 years old, a retired cab driver and airport shuttle driver, a lifelong bachelor. He is dressed fastidiously in a black double-breasted sport coat in double-knit polyester, gray trousers, pink button-down shirt, and bright red tie. He is in his church clothes, which he later says he bought from Goodwill. He is playing a tape he shot of a church choir singing, so he is free to move about the self-switching studio during the live weekly telecast. It is his responsibility to take care of all technical aspects of the show, including calling up the camera shots, turning on lights and microphones, and rolling in other videotape segments. The show moves slowly, with Johnny stopping frequently to tell his audience what he's doing technically. "All right now, I'm going to turn off my microphone and push this button here." After the choir tape ends, Johnny reads from a large piece of white paper on the desk before him, where he has seated himself. The words, in bold red Magic Marker, read: "I have some good days and bad days, Lord, but God has been good to me. Yes, He's been so good to me." He shifts from topic to topic, making announcements about local African American churches, leading his unseen, unheard audience in the Lord's Prayer, and, at one point, waving the observer over and announcing his "very special guest" in the studio that day. At the end of the show, he sings a gospel song a cappella. Johnny Dixon is one of a half-dozen

African American elders I met in Milwaukee who were pursuing their personal goals—motivated by their gospel beliefs—through participation in public-access television production.

African American elders are a substantial component of the so-called viewing audience.[1] What may come as a surprise is that some are *making* television, too. The five Milwaukeeans whose stories are related in this chapter demonstrate that a civic activism motivated by and filtered through the Christian Gospel, along with a charismatic personal style, has spurred a number of older African Americans to turn to public-access cable television to spread the word in their community about their favorite causes.

I started out probing the connection between African American seniors and public-access programming. The strong religious component that emerged here should not have come as a surprise, given the gospel tradition among black elders. Before long, however, I sensed that I had failed to anticipate the intrinsic presence of Christianity in the everyday lives of these five people, aged 56 to 84. This Christian element was to inform my whole approach, as it did the various television missions of the people I interviewed. I left interviews and cablecasts feeling I had been preached to. These people were all tremendous talkers, disarmingly charismatic and passionately convincing in their descriptions of their individual interests. In all cases, I observed strong ties between these oratorical and interpersonal skills and their traditional embrace of the Gospel. I saw these combined characteristics—along with several others—as coalescing to distinguish these research subjects from their peers, who may have watched much television but have never sought to appear on a program. Illustrations from them illuminate connections I have observed between African American elders, civic activism, religion, and cable production.[2]

African American Elders and Religion

Age is the strongest predictor of religiosity. In the United States, we often think of older people as more conservative and more religious, because people often are said to take on such tendencies as they enter late life. However, the religiosity of adults raised in the first half of this century is especially strong, owing to a period effect.[3] Many studies have suggested that Americans generally grow more religious as they enter old age and that heightened religiosity provides elders with coping tools and helps them achieve greater life satisfaction. Harold G. Koenig, one of the most prolific scholars on aging and religion, observes: "People value their religious beliefs and practices more strongly as they grow older because of a greater need to use them in adapting to ill health, finding meaning in life, and confronting their own mortality."[4]

Among elders, blacks especially are church members, attend church frequently, and perceive themselves as religious.[5] Religion traditionally has figured prominently in the lives of African Americans. Since slavery, the Christian church has represented a forum where blacks could combine worship and the promise of deliverance from oppression.[6] Elder blacks often look upon church involvement as the important work of lifelong learning or as an opportunity to perform roles with high prestige.[7] Black elders may command more admiration in their churches than whites in theirs because of what younger people perceive as the black elders' practical wisdom drawn from personal hardship.[8]

The complicated message that informs the consciousness of older black churchgoers comprises a gospel message of worship and praise intersected with church-generated discourses of civic awareness and political activity. These discourses shape black elders' knowledge about goings-on in their community and sharpen their awareness of power imbalances in society.[9] This all occurs in a forum that encourages them to glorify their Creator. It is relatively common for older black Americans to mix commentary on civic and political matters with praise for Jesus.

For the older blacks I met in connection with the work on this chapter, a southern heritage ambiguously shaped their worldview: having migrated to the North between the Great Depression and the 1950s, they still felt the sting of being brought up among an abused and despised minority in the South. (Of course, for many of them, they weren't treated all that much better up north.) However, as they reported, they felt gratified to have been reared in a region where family, community, and church played integral roles. Many spoke with regret that their own northern children and grandchildren had not had this "opportunity." For them, figuring their church prominently in their everyday lives was a way of maintaining connections with their (adopted) community, a way of maintaining the link between personal life and civic or political involvement.

For older blacks, their belief in the Gospel and the messages with which they link it—activism and political awareness in everyday life—may be the dominant sustaining influence in their lives. Largely without access to mainstream media outlets for their messages, some, logically, have turned to public-access cable to do the work they consider so important.

Public-access cable television offers a comfortable fit with the mainstream Protestant religious involvement of most older African Americans. Like the churches of these elderly, public access offers a forum for mutual concerns. Such a forum might be explicitly constituted by a gospel message, as it was with Brother Johnny Dixon's *Life after 60* show.[10] More likely, a gospel message shaped the discursive content of the programs involving the elders and helped contain their messages within the dominant religious discourse to some degree. Invariably, this was a message of hope and concern for one's less fortunate (often elderly) neighbors, who might need information related to civic life or who might suffer from circumstances brought about by structural inequality.

Older blacks may constitute one of the most prolific segments of public-access production because of the proximity of large portions of African Americans to the urban locations where the cable studios are located and because, as retirees, they can readily schedule time to use the resource. Most significant, however, may be the importance of their religious community. In announcing research about the primacy of religion in African American elders' lives, social scientists Jeffrey Levin, Robert Taylor, and Linda Chatters said:

> Black religious communities provide a context for understanding those special life conditions and stressors that are uniquely related to race and economic status. . . . The resiliency of African American religious traditions is found in their ability to confront these pernicious life conditions, to provide alternative methods . . . for their amelioration, and to invest diverse meaning . . . in those experiences. . . . African American religion, seen in these contexts, has produced a legacy of independent institutions and a tradition of worship that constitutes important spiritual, community, and cultural resources.[11]

The propensity for African American elders to relate their religion to a "legacy of independent institutions" may ultimately produce such a legacy for public-access television, which already bears the markings of democracy in ways that few other mass media forms do.

Cable-Access Programming and the Citizen

Public-access television started in the early 1970s, an outgrowth of federal legislation that allowed cities to trade monopoly cable franchises for public-, educational-, and government-access channel space, dedicated equipment and facilities, and training for citizen users. Barring commercial, obscene, or slanderous content, programming is uncensored, although, as public-access activist Brian Springer has pointed out, cable companies engage in "soft" censorship by actively failing to provide notice to communities that anyone is allowed to produce programming and have it distributed through these channels.[12]

Citizen-produced television has attracted much attention for three kinds of programming: the oddball, basement production antics, rife with antisocial content, represented in the *Saturday Night Live* sketch and feature film *Wayne's World*; the venomous telecasts of white supremacists and black antiwhite bigots; and shows that test the constitutional definitions of obscenity. These textual forms have not projected public-access as an outlet worthy of either audience or critical notice. But they do not represent the majority of public-access television; indeed, there *is* no majority—that is the point. And they do not represent the whole of its potential. In an interview with the *Humanist*, public-access activist Chris Hill illustrated this point:

> Even if 20 percent or 30 percent of it is stupid and boring, so what? Because in the end, I think that a lot of people don't understand how really limited is the range of information that they're currently getting from television. They might think they're getting all the different angles on any particular topic, but that's just a misconception they have from watching so much television.[13]

Religious programming abounds on public access. Milwaukee cable viewers, for instance, might see on two access channels shows on Muslim or the Nation of Islam, or they might see one of several Christian-focused programs, which, like other regularly scheduled shows, are kept in the schedule as long as the producers continue to provide them on a regular basis.[14] They are hammocked between sports talk shows and perspectives on the local arts scene, among several gay-activist shows, including the regular talk show *The Queer Program*, focused on younger viewers, and occasional programs about AIDS activism and informational programs provided by Senior Activists in a Gay Environment.

So far, public access has neither lived up to its potential as a major social liberator, as its developers may have envisioned, nor become a widely hated outlet. The chief reasons in both cases are that it has failed to attract large audiences and its schedule is filled with redundancies. What public access *has* done, however, is allowed pretty much anyone who wants to produce a message to do so and distribute it to all who might listen.

The measure of effects that public access has had on the kinds of messages our research informants have disseminated is questionable, to a point. Johnny Dixon's stated goal with his gospel show was to save lost souls. Did he do it? He didn't know—and neither do I. Sandra Lawrence, in producing *The Sandra Lawrence Show* with her 94-year-old mother as recurring guest, intended to save lost youth from crime and danger; I do not know that her program had a measurable effect. Douglas Jenkins and Charles Massey, in appearing on *Senior Forum*, intended to improve relations for blacks by publicizing the plight of former Negro Baseball League players and historical violence against blacks, respectively; I do not know that their work alleviated problems. One person who may have gotten direct results from public access was Yvonne Bennett, an activist among Milwaukee's black elderly who used public access's *Senior Forum* to get the word out in her community about how her longtime neighbors could hold on to their houses in a day when property taxes were skyrocketing as a result of white gentrification and the local media failed to publicize alternatives. Yvonne effectively educated her neighbors about reverse mortgages as a survival strategy.

The measure of success for public access in stories such as these is certainly difficult to discern. I do know that the stories these people told through it are oppositional to traditional, hegemonic frames that predominate on professionally produced television, where presentations of black speech and consciousness are often restrained by virtue of being made into modes of communication for a conventional medium. The accounts related here are stories with a clear point of view, one that may be in some ways radical (e.g., opposing white dominance) and in other ways conservative (e.g., Christian), but a point of view that does not fit the professional journalistic assumptions about the correctness of government institutions, the tacit racism in news reports that so often neglect the real circumstances of the black working class, and the phobic secularism that denies spirituality as a viable component of public discourse.

Introducing the Elders

Yvonne Bennett, 77, was a native Tennesseean who moved north after World War II. I went to see her after hearing that she was a local activist and self-described "idea person." She welcomed me and a graduate student into her north-side Milwaukee home for our first interview, where she presided at the dining table. Her house was old and somewhat in need of care, but she had it decorated with family pictures, plaques for her service, and other memorabilia. She served us bologna and spicy Buffalo wings on luncheon plates while sharing her enormous scrapbook, filled with news clippings about her civic work and such mementos as the city mayor's proclamation of "Yvonne Bennett Day" in Milwaukee.

Yvonne had suffered from severe health problems in her late years but believed the Lord's grace had kept her alive. She got around with some effort with a cane but also used a wheelchair. She wore dentures and, the first time we met, a gray wig, which migrated on her head when she scratched it. Yvonne had been divorced after a long marriage to a Baptist minister with whom she helped found a local congregation. She worked as a lower-level social worker for many years and matter-of-factly noted several times that color had been a barrier to her success. At 49, she had enrolled in college and got a bachelor's degree in interpersonal communication from Marquette University. In her retirement, she had made appearances on the public-access show *Senior Forum* show.

Before we completed our first visit to Yvonne's house, she required two things of us. One was that we walk to the back bedroom and look at her "office," a desk next to a bed, replete with an IBM Selectric and an expansive filing system. The other requirement is that we stand with her in her living room and recite the Lord's Prayer. Although we found this imbrication of academic and religious spheres shocking for our own part, we realized at once that we had rather asked for such an invitation by inviting Yvonne to talk about her public work for God. The more practical side of this was that we could not imagine anyone refusing this woman anything. (On our next visit, she talked the graduate student out of his pen.)[15]

Johnny Dixon, 75, was a Mississippian who came north during World War II and enlisted in the navy. He never married but said he had many affairs and had one daughter. He worked as a taxi and limousine driver, having left school in the ninth grade. He used to carry a revolver and said he spent much of his life angry, mostly at whites. He underwent a religious conversion about the time of his retirement and coincidentally learned to operate a video camera about that time. His show, *Life after 60*, was a combination of his preaching and his video clips from church services.

I found him to be an easygoing, gentle man. It was hard to believe his many stories about having been "feisty" before he was "saved." Johnny listened patiently to my questions and answered them thoughtfully and with vibrancy, standing up, for example, and punching air with his fists when he talked about getting into so many fights for so many years.

Charles Massey, 84, had, like Yvonne, appeared on *Senior Forum*. He had written a book, which was self-published, about his survival of a lynching in Indiana that killed two of his contemporaries. His "America's Black Holocaust Museum" in Milwaukee archived slavery and racial hatred in the country. These efforts were based on memories of an event that had occurred in his youth, an event from which he had never felt free. In old age, he felt as if he was finally coming to terms with his lifelong nightmare.

Massey took some classes at Wayne State University in his youth but was largely self-educated. Massey was a big man with a gentle smile and easygoing manner. He talked frequently of his love for family and his concern for other "children of God." I got the feeling that Massey wound up continuing his efforts—as proprietor of the museum, as a speaker to the public—in large part because he loved it, and that was as important to him as fighting hatred.

Douglas Jenkins, 61, had moved from the South to Chicago to play in the Negro Baseball League at age 19. After a career-ending injury, he became a college-educated social worker. Later in life, as the defunct Negro League grew in status through popular memory, he helped raise money for surviving players. He appeared on *Senior Forum*, and he was interviewed by many more traditional media outlets as well.

Jenkins understood the value of publicity for his cause. He solicited information several times about when the book would be published and who might read it. After our interviews, he stayed busy pressing his cause, meeting with Major League Baseball executives, including Milwaukee resident and interim commissioner Bud Selig, and appearing on news and talk shows, including *Oprah!* He felt gratified that his efforts had started to show results: the surviving players would get a pension for the first time.

I met Sandra Lawrence, executive director of a local youth project, at her office. She was boisterously busy in an environment she obviously loved and where she energized those in her service. Assistants flocked around her during our two-hour interview, catering to whatever need she expressed. I was entranced by her dynamism. Sandra was now 56 but had moved north with her mother in the early years of the civil rights movement. She was co-founder with her mother of the youth organization and involved herself in Washington-focused social service issues and events. She was host of public access's *The Sandra Lawrence Show*. Her mother made frequent guest appearances on her show to speak her mind on themes of her choice, usually moral or religious.

A graduate student and I made several attempts to interview Sandra's mother. I finally gave up after it became apparent that the 96-year-old woman had no intention of talking to us but was too polite to say so. I would have enjoyed seeing Sandra, who could talk at great length and who expressed such devotion for her mother, interact with her.

Religion: A Tapestry of Themes

I found that the elders involved in public-access programming shared several motivations for doing so. In a basic way, these motivations were attached to the elders' religious identity. They had made decisions to take certain kinds of messages forth for public consumption—generally, messages with a civic, secular theme—because of their gospel beliefs. For these five people, religion was absolutely fundamental. They made no harsh delineation between religion and everyday life and did not recoil from overt embrace of religion in the way that I am used to seeing religious middle-class whites do. They saw no humiliation in laying their claims of conviction. Indeed, everyday life for them was entirely embedded in their Christianity. They took seriously, then, their role as crusaders, as witnesses. Even if their public addresses rarely mentioned God, except for Dixon's show, they considered themselves to be spilling forth the Gospel. Their television work constituted an expressive means of conveying this gospel belief, and it allowed them to connect with their belief in an afterlife, or other-worldly life. Additionally, this television work, like their religious affiliation, awarded them a special status in their community, a quality that seemed important to many of these elders. Finally, the work these people did seemed motivated by their quest for meaning in life on earth.[16]

Television as Expression

For Yvonne Bennett, an outlet for expression of her ideas is vital—I halfway believe that without one she would die. "They call me the idea lady," she said numerous times during an interview at her home. As I pored through her treasured file of community write-ups, letters from local dignitaries, and so forth, her "idea lady" self-description surfaced several more times. Yvonne was no stranger to media prior to her public-access use of television. Before she retired, for several years she had hosted a local AM radio show called "Good News You Can Use," wherein she contributed her "ideas" to local seniors from her vantage point as a senior social worker. At one point Yvonne commented on her files of papers: "You know, you may want to write a book [about these]. I mean, help me write a book. See, I'm fixing to write a book."

During the visit, Bennett readily supplied dozens of anecdotes about her one-woman battle against the system to let common people know their rights and get their due. She clearly enjoyed this crusader role and relished telling about it. She had approached Worthington Hortmann, *Senior Forum*'s host, about doing his show after the state government had cut off funding to let seniors know how to deal with the reality of their rising property taxes. She recalled: "Mr. Worthington and me, we went to Madison with the video camera, and I delivered a petition with signatures and stuff to replace the funding. . . . We had video showing that we made sure our application got in first, because money was so short. We took a picture of the place where we went and broadcast it on the *Senior Forum* show." Clearly, Yvonne understood that she wanted to use television to express her ideas, with the emphasis on change.

For Johnny Dixon, the expressive function of public-access production worked differently but still was essential. Dixon told us how he was always playing around with a video camera, mostly at church and family functions, and how he had loved to use his VCR to make tapes for people to watch. Then a friend suggested that he go down to the cable authority and take some training for fifty dollars. He jumped at the chance to use his video-making hobby as a public outlet, although it caused him immediate discomfort:

> I was kinda scared when I came down here and first got on TV with *Life after Sixty*. I don't want to be on no TV, 'cause I was always shy, y'know. And so, I started to . . . I did one show and I got confidence. Get more confi-

dence as I go . . . I want to show my gospel tapes to reach out and try to touch some people that maybe won't even talk to me. Like, you know, maybe some people wouldn't even let me in the house. This way, they might watch the program and it'll make them feel good. I like to just put out good stuff to make them feel good.

Dixon's self-described shyness was not apparent to me, given his loquaciousness. He talked at length about many topics, including his checkered past and his pride in the suits he wore. Troubled for many years and plagued by a bad temper, Dixon had been "saved" in the Pentecostal tradition in his 60s: "I want to tell people about it," he said. "I want to save some people from what I went through." When we met, he had been doing the show for about one year and often spoke to his audience of "the Lord's goodness" but had not yet disclosed his own sinful background. "I'm going to get to that," he explained to us, but he wanted to gain his audience's trust first.

Every two weeks, he produced *Life after 60* live, by himself, using tapes from his mountainous collection of choir videos, the quality of which was spotty, limited by his real-time, one-camera production efforts and uncontrolled location shoots. Each show appeared in rerun the following week. Dixon hoped that his message was getting through to many people, but he admitted that he had little idea who his audience was, aside from the occasional "sister" at church who might approvingly remark that she had seen the weekly program. Gesturing toward the studio phone, he told us that he was going to start a live call-in period that would help him reach out more directly to his audience.

For Charles Massey, publicizing the plight of African Americans was urged by an expressive impulse. He explained:

I got this idea when my wife and I was traveling over in Israel following the path of Christ. . . . I had the occasion to go into the Jewish Holocaust Museum in Jerusalem. . . . I was emotionally disturbed. I came out of there and composed myself. I told my wife, I said, "Honey, we need a museum like this in America to show what has happened to us black folks and the freedom-loving white people who've been trying to help us ever since we been in this country." So this is it. This is the only black museum of its kind in the world.

As for television, he saw it as an outlet for his own public expression, even if he condemned it as a medium for being a "damn idiot box" lacking in programs that were "uplifting or moralistic." He noted, "It's ruining the whole moral fabric of our country. . . . Television is the ruination of our youth. They show them shooting dope and killing people and all that stuff. So people get a gun and do what they see somebody on television do. It gives all kinds of ideas to them." Massey had no trouble identifying what he thought was the reason for television's being this way: "to maximize profits." People such as himself, he said, had to work to reeducate people to be more critical, more discerning; he sensed his role as civic leader and saw public access as an outlet for expressing this role. And that is why, he said, he took the news of his museum both to Hortmann's *Senior Forum* show on public access and to other outlets: "It's important to remind people," he said.

Television Work as an "Other-Worldly" Connection

All these elders spoke of an other-worldly impetus guiding them in their efforts to reach others through public-access television. For most, there seemed to be a direct connection between God's plan for them and their television work. For example, Charles Massey commented on his television appearance in which he related his personal experience of having been the victim of a lynching party as a youth:

> So they . . . they lynched me. . . . But they didn't lynch me with a rope around my neck like they did Tommy and Abe. They lynched me with a severe beating, although they had the rope around my neck. And it was a mysterious voice [among the crowd] that saved me. It was a voice from Heaven, that's what I believe. God. He had something for me to do. I hope I'm doing what He wants me to do.

Likewise, Douglas Jenkins said his efforts to publicize the Negro Baseball League players' lives through public-access television were an extension of his spiritual beliefs: "I don't get to church like I should anymore, but I believe in the things I'm doing and that the Lord is sending me. He's doing it through me."

Sandra Lawrence connected her media work to a sense of duty to God: "I have done what God has sent me to do. At the end of life, what matters is that you love everybody in the world, that you do what you're supposed to do. And I feel like every morning I get up and do that."

Status

Sandra Lawrence talked at length about her status in her community and on the national stage as well. It was clear that her program, named for herself—*The Sandra Lawrence Show*—was in part about achieving status in the white-dominated world, in a way that is similar to the means provided by her religiosity. Early in the interview, for example, Sandra remarked that the president of the United States had invited her to the White House to seek advice on the crime bill, and she had gone and "shared with him." She explained: "I'm a community leader with a TV show—people call me about any issue, nationwide. . . . I've had Jesse Jackson on, Farrakhan on. I know these people personally, right? The national shows and CNN will call and ask, 'Sandra, can you put me in touch with his person?' . . . When Jesse Jackson was in Chicago with six people seated at his table, I was one of those six people." Lawrence seemed unconcerned with viewership numbers. She was host-focused: "People know that to be on my TV show, it's good for them. If you're a politician, it's really good for you." She noted that her show had won "a lot of awards" and had been bumped up to the "big studio," with a live audience that could be pictured in the telecast.

Associating her name with guests on the program seemed important, as similar associations were to Lawrence: "I am really loved. I am really loved [in the community]. I'm at a level like Dr. King." Later, she said a conservative Washington foundation had awarded its special honor to only four Americans: Bob Hope, Barbara Mandrell, Lee Iacocca, and herself. As for television, Lawrence seemed to see her status role as a link between individual empowerment and civic responsibility.

I wondered if Lawrence's concern for her own status was tied up in her perception of her mother as a great civil rights leader. She said she had always been close to her mother, who had brought Sandra with her across the country during her campaign for civil rights in the movement's early years. Sandra had many great stories to tell about her mother's efforts, including being bitten by police dogs in Selma and preceding King in the movement. She told a charming anecdote about a Bill Clinton aide's having called Sandra at home to ask her to appear locally at an event that evening; her mother answered the caller: "She might go, but first she'll have to come home and make her bed. She didn't do it this morning." Lawrence seemed lovingly aware of and deferent toward her mother's apparent status but, perhaps because of it, was mindful of her own. And, as with several other informants, when she occasionally became mindful that she was perhaps saying too much about her own status, she invoked God: "Of course, this sounds so much about me, but I do it all for the Lord." And later: "What has kept the agency going for twenty-six years has been my divine inspiration from God and belief in him." And again: "You have to understand who I am. People know me from television here and from being on national TV, OK? Who I am is a person who created an organization that saves children's lives, saves family's lives. That's who I am. The reason I can deal with all those things twenty-four hours a day is because I do love God. I allow him to live in me."

It reminded me of Yvonne Bennett's half-apologetic remark as she related her long story of civic activism: "This all may sound conceited, but I do it for the glory of God." Bennett's own talk consisted in part of her describing her journey of activism as an important performer. For example, when asked about her marital life, she spoke little about her first marriage to a laborer but described her second marriage in terms of the status it brought her: "I was married to a minister. He built the first black church, first black Baptist church [in Milwaukee]. And it was all built by blacks." Later, she spoke in similar terms about her career and civic activism. For example: "I can tell you so much about discrimination and how I wrote proposals. I was active in the movement, too. And, 'cause being one of the first ones here, you know, I was like a pioneer."

With her experience doing a radio program, it seemed natural, Yvonne said, that she should join Worthington on his show several times to promote her case for elderly Milwaukeeans learning about reverse mortgages in order to preserve their homes. "I've always been a motivator," she explained. I believed her.

Status, in light of the discrimination his race has faced, also informed Charles Massey's perspective about appearing on public-access television. Massey, however, tacitly recognized the respectability associated with more professional, "middle-class" media outlets, something that was absent from the speech of some of the other informants, such as Dixon. When Massey was asked what sort of television he had been involved in, he did not mention public access immediately: "I've been on NBC, ABC, 60 Minutes, [*Jerry*] *Springer*, *Larry King*, *Oprah Winfrey*, *Tony Brown's Journal . . . The 700 Club* with Pat Robertson, all the television stations here in Milwaukee, even the state, public television in Madison. People from Holland came over here. The British Broadcasting Company came over here and made a forty-eight-minute documentary on me." The work of publicizing his museum—and the work of the museum itself—was a job that Massey said he did because God wanted him to. "I'm enlightening so many people," he explained. He told a story about doing a location shoot with CBS at the spot of the lynching tree where his friends had been hanged: "And a young white girl came driving a car and she wondered what was going on. And she found out what was going on—that I was Charles Massey, one of the survivors of the lynchings. And she ran and parked her car and then she ran up while the camera was going and said, "Oh, Mr. Massey, I'm so sorry about what happened. I'm so glad God spared you." For Charles Massey and others, personal status was an indicator of salvation. This link was a marker of working-class identification (if not current working-class status for all these elders); to be linked with salvation meant respectability.[17] As Robinson noted at one point, "Whatever you do, if you do it for the glory of God, he will take care of you." Without fail, these people gave some version of this explanation. "The Lord is guiding me," Douglas Jenkins said about his efforts to publicize Negro League baseball.

Meaning

Douglas Jenkins, asked why he took his message about the Negro League to public-access television, connected his effort to the broader meaning of television in the everyday lives of young blacks:

> The kids I [am talking to] are not your typical TV kids. These are the kids that are in the streets all the time. Nobody's at home for them. They could probably get the story before the newscaster gets it. That would be great if their mama knew they were home watching TV. They're out there doing things that they shouldn't do out there. . . . They are constantly involved with some kind of real violence, whether it's sitting there watching it, knowing about it, or being involved in it. This is an everyday occurrence. They're not afraid of death. . . . TV is nothing like what they see.

Jenkins said he wanted his own story—the story of the great Negro League players who can now be seen as American heroes—to be told for such young people, whose behavior and understanding of life can be helped by it.

When Charles Massey explained why he devoted efforts to publicize what he called the Black Holocaust, he revealed his quest for meaning: "Have you ever heard the story of Ida B. Wells? She's a black lady that fought against lynchings all the days of her life. She was burned out, shot at, maligned and everything else. And she never had the chance to put up a Black Holocaust museum. I consider myself taking up where she left off." It seemed evident that the search for meaning was a religious endeavor for Massey, who said he found his adopted Catholicism empty in its rituals contrasted with the African Methodist Episcopal Church of his childhood: "I wish they would have more clapping and shouting and things like that in their program. But I can still get by and feel the spirit of God moving with me. But the thing is, black people are used to making a joyful noise to God."

Most of the people I interviewed, at some point or another, crystallized their endeavors through public access in a few words that portrayed their efforts as quests for meaning. For example, Yvonne Bennett noted, "I do it because I might be helpful. . . . I just want to be able to motivate black people. To help them use that creative mind."

Johnny Dixon remarked: "The reason I do it is because I was out in the world all my life, unhappy. And now, for the first time in my life, I'm happy within, you know? When I was out there I was like a live bomb walking, ready to explode at any time. . . . I'm being blessed."

Sandra Lawrence said her show's purpose was to distribute information to people who needed it. "Someone like me, with thirty years of being out there, knows how to use the media. You learn to think, 'How can I best serve the interest of the people?' " As Lawrence makes evident, some functions clearly are connected. Lawrence's search for meaning articulates with her concern with status: "I'm the one they can reach out and touch. Like, if I walk down the street, you would think I'm Mother Teresa. That's why they want to touch my hand and hold me. I'm somebody who would take time to hear about their little kids. So it makes a difference. And they rush to their TV sets to see me."

Service, not money, is her greater concern, Lawrence said. (This is a theme I heard echoed by others). For both women, it was important to acknowledge that greed was an improper motivator for seeking public attention but that there were right motivators, and these were God-based. In Lawrence's words, her television show is about possibilities, not limitations:

"Now, what stops a lot of things is that people think money. I don't think money. If I see something, I go do it." For Sandra, public access made television publicity *possible* for her cause. She and Yvonne both talked about their desire to use public access as one of several outlets of the media that they could capitalize on. Yvonne, for example, had had many experiences with publicity flyers, local radio, and the black press in Milwaukee. Public access, for her, was another means to accomplish her goal, whatever it might be at the time. Public access seemed to differ from other "black" media outlets for Yvonne in that it offered her an unfiltered voice; she spoke of not having to conform her message to fit tight formats in forums controlled by others.

Watching Some of the Programs

The program *Senior Forum* opens with a three-minute, laboring montage of preproduced credits and shots depicting "active" and smiling elders eating, dancing, playing cards, watching television, and doing crafts at Bethesda Senior Center. The Count Basie song "Take the A Train" plays in the background, and an unidentified woman, who sounds to be in her 70s, reads a long voice-over as the words crawl across the screen, superimposed on the montage. Appearing frequently in the shots is Worthington Hortmann Jr., the senior center's director and the show's host, who is in his late 30s. Hortmann does the live portion of the program from the cable-access authority's "impact" studio, where he generally performs the self-switching tasks between two fixed video cameras, one of which shows the "desk" set head-on and another of which is set to display a close-up of some artifact that might be placed atop the desk, such as a photograph or a computer-generated printout of a community event bulletin. With the head-on camera, Hortmann is able to switch from a single shot of himself, as host, to a two-shot, wherein he is able to capture the conversation between host and guest. Like Johnny Dixon's production efforts, which are detailed to some extent at the opening of this chapter, Hortmann's work is somewhat stilted. At the end of the opening credits, Hortmann welcomes viewers and assumes the character of "Grampa." The character is not an obvious persona—only a nickname. Hortmann does not act any differently when he is Grampa than when he is not, when the show is not being produced.

On the day when Douglas Jenkins was his guest, Hortmann announced Jenkins's name and went on to perform a series of "shout outs," wherein he greeted, on the air, such people as "Reverend Buck and the Milwaukee Bucks, the Pittsburgh Hortmanns (his visiting relatives), and Ethel Anderson, who watches every day." Soon, Hortmann begins a spontaneous interview of Jenkins, who first is asked to give his own "shout out" and does so to his family, while fumbling to recall his new daughter-in-law's name. Next, Hortmann asks Jenkins to relate his story of the Negro Baseball League, and he asks a few questions, punctuated by "And, uh," and "OK," which help to belie the program's status as professionally produced television. Soon, Jenkins has launched into his story by way of answering Hortmann's questions, and he recalls infrequently to look directly at the camera. He pauses and hesitates often and follows tangents in the stories he tells, but his stories seem earnestly told, with a certain amount of critical distance.

Jenkins portrays himself as insider/outsider: he often talks of his membership in the Negro League but talks about the players, most of whom had joined the game when he was a child, as historical figures. He talks for long stretches, filling most of the half hour with anecdotes about league luminaries such as Rube Foster, the league organizer, and Jackie Robinson. All of these anecdotes combined to produce a moral tale of good Negro League players who did not get what they deserved from the white system, and Jenkins frequently mentioned that his own work was to publicize this story so that at least some social recompense might be paid these men, most of whom had been older than Jenkins and were now dead. Lots of Negro League players had been ready to play in the major leagues before baseball owners stood up to the prejudice of fans, players, and fellow owners, Jenkins said. Brooklyn Dodgers owner Branch Rickey selected Robinson in 1945 to break baseball's color barrier, Jenkins said, "not because of his superior playing ability but because of his attitude." Robinson, Jenkins said, was signed because he was an educated black who knew how to "handle the press" in a racially divided nation. Hortmann, for his part, expressed fascination at Jenkins's revisionist tale, occasionally looking away from his speaking guest toward the camera to make such intonations as "Did you guys know that?"

Hortmann's guests, for the most part, including Jenkins, addressed him rather than their "audience," but Hortmann worked hard (through his direct address of "you guys" and his "shout outs") to keep his audience with him, whomever it might have consisted of. While his guests maintained a serious air, Hortmann smiled, laughed, and expressed concern visibly when it seemed appropriate to him. He knew he was the link between the storytellers and the "live" reception of their tales. With time running out, Hortmann thanked Jenkins for appearing. Although most of Hortmann's guests, such as Massey and Bennett, made frequent allusions to their religiosity (Bennett told her audience directly, "The Lord told me to come here, and I did not want to be disobedient!"), Jenkins did not invoke God in his interview responses on *Senior Forum*. Later, when he explained his work for this book, however, his religion figured large in the discussion, as he volunteered that he was spiritually motivated and he was concerned with God's love of black people. (Hortmann did make mention of God on the Jenkins installment of *Senior Forum*, closing his show by pointing his finger at the camera and admonishing viewers, "Now, go to church this Sunday!")

A different sort of public-access project was *The Sandra Lawrence Show*, shot live before a studio audience of twenty-five to thirty people, mostly African Americans, who collected in the studio from the community. Before the host, Lawrence, would appear, the show's high production values became evident in the studio, as light and sound technicians appeared to perform technical preparations and a young man took the stage to "warm up the audience" by telling jokes and to instruct the audience in the performance of its duties, such as applauding on cue. A minute before airtime on one particular night, Sandra and three guests, two social workers and a teacher (all white), took their places on the set and got their microphones attached and tested. Then the studio monitors displayed the prepackaged opening, which was now airing: up-tempo music accompanied a flossy montage of the activities of Lawrence's youth center program. The sequence dissolved into a live shot of the studio set, where Sandra, a large woman who dresses vibrantly, was shown in a medium close-up, smiling graciously at her applauding audience: "All glory to God! I'm Sandra Lawrence, and welcome to *The Sandra Lawrence Show*!" She began a long "prayerful open," in which she sermonized without notes or other prompts, with her eyes closed and hands raised. Next, she introduced a young woman who sang an African hymn a cappella. After applause for the singer, Lawrence shifted attention to the interview set, where she sat at her desk next to the three visitors. She led them in a discussion of her youth program, educating the audience about the center's features (such as a drug treatment center) and the coming black-tie awards night.

Near the show's half-hour mark, Sandra introduced her mother, a former civil rights leader, and asked her to speak. The 94-year-old woman spoke in a rambling fashion at first but then spoke clearly and in a captivating tone, advising her audience with such directions as: "Quit your selfishness! Trust in God!" After her mother's appearance, Lawrence introduced and interviewed three other center employees, concluding, "At the center, our services are free. God and the community have put it there for you." Finally, the singer took the stage again for another a cappella hymn, "The Greatest Love," and Lawrence reminded viewers about the center's awards banquet.

All during the telecast, Lawrence appeared focused, in control, with an electrifying presence. Unlike Hortmann and some of his guests, who had trouble traversing the worlds of interpersonal communication and television production, Lawrence moved fluidly from topic to topic and deftly split her attention between guest, studio, and camera (the surrogate home audience). She never fumbled for words, either in her prayers or in her interview questions, and her agenda—publicizing the center and assuming the cloak of the Gospel—was never far from the surface. Her voice had a calming, resonant quality, and she talked quickly—as she had in our interviews—constantly sprinkling her main points with anecdotal evidence about the goodness of humankind and the compassion of Jesus. Her considerable polish was not without assistance. Ritually, Sandra quietly played meditation tapes during her performance—featuring sounds of the ocean or perhaps a light rain on a tin rooftop—in a tape player that she kept on a table next to her interview chair. For Sandra, these quiet reminders of God's power and glory were means of maintaining her own steady strength.

Conclusion: The Public-Access Crossroads

The people who participated in this study shared some important attributes. They seemed to be expressive, strong-willed, highly committed individuals, dedicated in their service to their Creator and in their compassion for others, especially other members of their race who, they felt, needed their help. All were charismatic leaders. Yvonne Bennett, for example, received phone calls and visits during her interview from admiring elderly friends, who looked to her for guidance. Sandra introduced me to an office staff that stood in evident awe of her, seeming to delight in answering her every need. And I left each interview enchanted with these inexhaustible individuals.[18] Each seemed to have a burning need to impart a particular message; often, it was a message having to do with inegalitarian social structures involving race and class. (Interestingly, the two women in this study never raised gender relations as an issue.) Although only one of these individuals' issues primarily was saving souls (Dixon), all of them delivered Christianity as a strong subtext. They conveyed their messages with the same sense of urgency, of comfort, of community that we might expect to hear from the pulpit. All of them, in fact, saw their public-access efforts as their own personal ministries—attempts to help people in the name of the Lord.

Some, such as Yvonne Bennett, generally were television fans; others, such as Charles Massey, derided the medium on the whole. In fact, like most older people, all leaned toward news and public-affairs genres for their own viewing preferences.[19] But all saw public access as a means for them to use for the public good a medium that they recognized as powerful because of its pervasiveness in their community (even if their knowledge of the limitations of cable subscription among seniors was a difficult truth for them to reconcile). They typically acknowledged the place of public access outside mainstream television. Johnny Dixon, for instance, reported that he knew that his production competence was not as great as that of a professional, and he was able to tell us how it had improved since he made some of his earlier videos. Bennett said she was glad public access existed to supplement "the white media." She explained, "They aren't always interested in explaining things to people, especially the elderly. We can do that through the access cable." She told us that professional news outlets were less interested in issues that took a long time to explain and that may affect only older citizens. Sandra Lawrence saw her public-access show as a means for reporters to pick up news items and place them in the "mass media." She noted, "I do realize they work within a framework." But sometimes, Sandra said, she had been able to influence black reporters in "the white media" with stories about her agency, and some stories spread from her show to reporters who learned about the coverage.

It is easy enough for me to relate impressions of these elders—that they were self-starters, opinionated, dedicated, and so forth. But I must acknowledge the limitations I brought to this impression-forming context. I shared a background with the elders but important differences existed. I—as was the graduate student who helped conduct interviews—was white and part of the upper-middle-class world of academics, whereas all the elders were African American. I was a couple of generations younger than the elders. I grew up in a working-class family in the rural South, as most of the elders did, and in the religion of southern Protestantism, which held some resonance with the purely religious views expressed by the African Americans. At the time of the interviews, I lived in an affluent suburb of highly segregated Milwaukee, but the people I interviewed lived in the primarily working-class area of the highly segregated city where most African Americans reside. Three of the elders had college degrees and shared with me a movement into the middle class as adults, one was largely widely read and self-educated, and another had no college experience. I wondered how our differences, particularly racial differences, bore on the elders' stated views. For example, I wondered whether they might have, on the whole, more readily criticized whites if their questioners had not been white. And, while I believed our differences kept us from taking salient points for granted, I also wondered if I might have been able to tell a tale of greater understanding if we shared more in common.

My curiosity increased when I discovered a discrepancy in the story that Charles Massey presented for this book and the account that he rendered on Worthington Hortmann's *Senior Forum*. I noticed that the two accounts were so similar as to have appeared practiced; in fact, all the elders' interviews contained passages that were virtually repeated from speech they had elicited in the telecasts in which they had participated. It seemed evident that they had delivered their message so many times that they had boiled it down to a memorable narrative, replete with favorite key expressions. In Massey's case, in our interview, as noted earlier in this chapter, he explained the genesis of his black holocaust museum this way: "I told my wife, I said, 'Honey, we need a museum like this in America to show what has happened to us black folks and the freedom-loving white people who've been trying to help us ever since we been in this country." And he left the statement at that, although he did go into his concerns about the considerable maltreatment and oppression of blacks in the history of the United States. But, crucially, he did not saddle his white interviewer with blame by lashing out at whites more provocatively. In his appearance on *Senior Forum*, which had taken place several months before the interview, he made virtually the same statement about telling his wife he wanted to establish the museum in behalf of "us black folks and the freedom-loving white people." Then, seated next to his African

American interviewer—and, I believe, assuming a predominantly African American audience—Massey continued:

> They have a group of people that came over here, call themselves the Anglo-Saxons. They consider themselves God's gift to humanity and which they don't want to give people their due credit for anything in this country. The Jews, Polish, the Germans, the Soviets, the blacks, and the Hispanics all have contributed as much if not more to our civilization and culture than these folks called Anglo-Saxons, who pride themselves on being God's gift to humanity. But if God had consulted me, Mr. Worthington, before he created these Anglo-Saxons, I could have saved him a lot of trouble. [Hortmann smiles.] I could have saved him a lot of trouble. You see, the thing of it is, they have been ramming this down our throats from time immemorial.

A similar occurrence happened during my interview with Sandra Lawrence. As we got acquainted, I neglected to mention to Sandra that I, like her, was a southerner. As our interview progressed, a mounting sense of guilt plagued me for not having remembered to do so; Sandra proceeded to censure broadly and sharply white southerners for having made life miserable for her mother, and, to an extent, for herself, as her mother had battled for the civil rights cause, especially in Selma. At the same time, I realized that, if I had remembered to mention my own origin, Sandra probably would have tempered her words. When our talk was done, Sandra gave me a hug and happened to ask where I was from. "Alabama," I said, and she let out an audible gasp, then a smile. "Well, God bless you!" she told me. Significantly, both Lawrence and Massey had been as conscious that they had an audience during a scholarly interview as they had been when their words were being taped for television, and they edited themselves accordingly. It is a shortcoming of this book that, in my aim to uncover rich diversity, I have constructed so many "realist tales" from politically charged interviews that I have no doubt misunderstood many times.[20]

What I felt I was able to relate from the conversational exchanges with these five older African Americans, though, was that they saw a great stake in the use of public-access television, although none of them seemed to dwell on the role of the audience. Proponents and critics of public access variously have characterized this outlet as guerrilla, subversive, participatory, and a soapbox, and I can argue that the elders discussed here have used all these applications. The forum's place, ironically, has been ensured by the very government that often gets criticized through it. That is partly because it does not cost taxpayers (except in their roles, indirectly, as cable subscribers). The other reason, as legal scholar Jason Roberts has observed, is that decision makers realize that "public access has been the medium for those whose messages cannot be heard elsewhere."[21] Indeed, it is also a medium for those who are, perhaps because of age, unable to leave their homes to engage in unmediated social discourse. Through Brother Johnny, they can "participate" in church services that are so meaningful to them or through Yvonne Bennett learn whom to call in order to avoid being turned out of their homes.

However, public access is imperfect as a democratic instrument, in much the same way that Thomas Jefferson realized that freedom of the press was an ideal that did not work as a practical matter. Jefferson pronounced press freedom as vital to society only if all citizens had equal access to news channels. Unfortunately, many of the people whom our informants wished to reach with their public-access messages could not receive them. They could not afford the twenty-eight-dollar hookup fee and twenty dollars–plus per month basic cable subscription bill.[22]

Some theorists have argued for the kind of egalitarian social landscape that Jürgen Habermas desires, a condition that Benjamin Barber has identified as "strong democracy."[23] In a strong democracy, all people have free access to means for affecting their basic social circumstances, an ideal that grows murky when we consider the representative form of government in the United States and most other Western societies. Richard Sclove tempers Barber's idealism when he acknowledges the possibility of engaging in a partial strong democracy without radically restructuring the basic elements of U.S. government. Fundamental to this democracy, Sclove urges, is a technological system that encourages the free expression of ideas along such lines as class, gender, and race—as well as the opportunity for civic activism to take place in real public spaces, not just virtual communities: "Technologies should—independent of their diverse focal purposes—structurally support the social and institutional conditions necessary to establish and maintain strong democracy itself."[24] Sclove goes on to point out that one reason television cannot replace human gathering sites as the center of the public sphere is that the medium filters out the "nuances" of human experience.

More important, perhaps, for Sclove is television's "potential hegemony of experience"—the power of the audience, in effect, is unequal to that of the producer.[25] This is no less true when the television producer operates from a worldview that essentially matches that of the audience, as it must for many public-access producers and their special-interest viewers.

What do we make of television following the introduction of cable-access channels? Do such channels, with their surprisingly large range of production standards and paltry-sized audiences, even *count* as television? I contend that, although cable access has not had and may never have a revolutionary effect on the medium's dominant forms and content, it has begun to carve out an important niche. Serving neglected audiences, championing unadulterated interaction, and disseminating diverse messages, cable access could constitute the single most democratic influence on television since the medium's invention.

In fact, on one level, the uses these elders are making of public access provide something of a remedy for the lack of public dialogue in contemporary life that has been decried by Habermas. As I have noted in chapter 1, opportunities for nondominant others to be heard publicly are rare. For this reason, social spaces such as public access, where these elders have aired their views, facilitate autonomous expression. As philosopher Nancy Fraser has put it in her counterargument to Habermas's idealized public sphere, stratified societies such as our own prohibit full parity of participation by their citizens, and when a single deliberative space exists as the public sphere, it inherently wreaks havoc on the interests of marginalized people. Fraser is heartened by the ad hoc appearances of what she calls "subaltern counterpublics," wherein members of a particular subordinated social group have come together to constitute a "discursive arena" through such expressive outlets as research centers, festivals, and video distribution networks. Issuing the caveat that subaltern counterpublics imply no intrinsic virtue, Fraser celebrates the practical potential for these publics to "expand discursive space." Citing the measured successes of the feminist movement as her key example, Fraser notes that subaltern counterpublics are able to move what many people consider to be private (publicly taboo) matters, such as domestic abuse, onto a public agenda for deliberation; such is the exercise of "publicity"—making the private public—in the least pejorative sense.[26] As elders such as those interviewed for this chapter continue to work in public-access production, perhaps they will contribute materially to the efforts of various "subaltern counterpublics" to which they may belong. Perhaps, as Sandra Robinson hopes, her place within a grassroots movement to help direct the success of young, urban black people in a society that shows little regard for them will not only help make her cause more real through its publicity but will result in beneficial, if incremental, change (as Fraser has argued about the women's movement).

But public access, even used in this vitalizing way, counterpoises its own benefits, presenting a drag on democracy as we know it. If the identity politics being preached by such groups as the elders I've described here is carried to too great a length—and I'm not sure we can identify what such a length might be—cable access could have a disruptive influence on the body politic. As Alexander Meiklejohn argued more than thirty years ago, the constitutional guarantee of freedom of speech and expression is inherently paradoxical: the First Amendment explicitly prohibits any abridgment of this right, but for a society to operate at all practically, this right must be abridged at many turns. For example, Meiklejohn notes, it is against the law to shout "Fire!" in a crowded theater; such a prohibition is entirely defensible, because this "freedom of speech" would endanger public safety. The First Amendment, Meiklejohn insists, is not the "guardian of unregulated talkativeness."[27] Crucially, for Meiklejohn, the First Amendment necessarily means that the opportunity for everyone to talk carries less significance than the need for everything that is in *the public interest* to get said. Agreeing with Meiklejohn, media scholars Theodore Glasser and Stephanie Craft have pointed to problems with the so-called public journalism, a still unfolding model of news distribution whose proponents have so far failed to come to a consensus about how to define public dialogue. Glasser and Craft are careful to say that democracy depends on deliberative work, not on the volume of talk. As they note:

> Except perhaps in the smallest of communities, and excluding the unproven potential for truly democratic participation in cyberspace, "direct" democracy has been difficult to achieve and even more difficult to sustain over time. The problem is largely, though not entirely, one of scale: Where do we find, or how do we create, democratic associations small enough to accommodate full participation among its members?[28]

Notwithstanding Glasser and Craft's cautious optimism on cyberspace, this problem of establishing and maintaining a forum through which all voices might be channeled is structurally similar to the problem with the Internet: many voices speak, but no clear criteria exist to judge their credibility and what should be their level of authority. In such a relativistic atmosphere, abetted by a fairly unfettered set of technologies, governance may begin to lead not toward a greater degree of democracy but toward anarchy. This conundrum—of whether an imperfect, uncensored proliferation of voices might further weaken a political system that currently is afflicted by a restriction of participation—is the riddle we are living with at the opening of the new century. Where people's creative use of such technologies as the public access will take us on this front will no doubt remain contestable for some time to come.

Five

A Pinch of Salt

In 1988, the white actor Richard Crenna starred in a workaday network miniseries called *Internal Affairs*. Crenna plays Frank Janek, the boss of a special investigative team of the New York City Police Department. He is a crusty, sixtyish rogue hero much revered by a staff of detectives from various ethnic backgrounds, with a woman among them. Janek is Dirty Harry for television, the bruised white knight complete with a youthful love interest and damsel in distress played by Kate Capshaw. One elderly person has a speaking part in the show, and she is only on camera for a brief scene. She is the former madam of a Saigon whorehouse and appears to be in her 70s. In the story, before we meet her, Janek's protégé has flown to Tampa to interrogate her in her mansion, where she is living the grand life of the most successful Sun Belt retiree. The burly protégé calls Janek and tells him the woman is too tough for him to get information from and that it will be necessary for the senior cop to join him for an interview. "What's she like?" Janek wants to know. The protégé answers coyly: "She's like your Chinese grandmother would be if your Chinese grandmother were really evil."

The retired madam lives up to this baleful description, demonstrating more concern for her tee time than for Janek's questions about the unsolved death of a prostitute who had worked for her years ago and the sexual abuse of children in her competitors' employ. She speaks in mean-spirited, clipped tones. Surrounded by her subtropical trappings, limousine, and house staff, she is country-club classic. Her long, black hair is swept up in a conservative twist, framed by a crisp, white visor, complementing her pastel golfing attire. This petite, elderly woman has achieved the American dream but in a perverted sense: we know her not just as a rich, old woman but as a former Vietnamese bordello keeper who looked the other way when GIs paid her for the privilege of hurting young prostitutes.[1]

Presentation of an elderly Asian woman on network entertainment television is a rarity. This miniseries character's embodiment of nearly undiluted evil may have seemed a sad stereotype but was a petty digression from television's practice of ignoring minority elders. For "mainstream" consumption, this older woman came across as already encrusted with deviance. There was a salacious value to her character. She contributed a quirky, nonwhite counterpoint to the "normal" picture of old age, as represented in Janek. We were, of course, meant to identify with the cop.

Other than a scant complement of African Americans, mostly men (Redd Foxx, Ossie Davis, James Earl Jones, for example), it is difficult to name elderly members of minority groups who have enjoyed broad exposure on contemporary American television. Elderly homosexuals, in fact, seem to be categorically shunned, as are older members of most ethnic groups.[2] If American popular culture downplays the existence of elders generally, it virtually ignores those who might not be heterosexual whites. In a study of television advertising, for example, social scientists Abhik Roy and Jake Harwood found that not only were elderly women represented in commercials much less often than elderly men, but African American men were portrayed rarely and elders from other minority groups were entirely absent.[3] Yet elderly members of minority groups have television sets and are known to buy many of the products that are marketed through their advertising system.[4] What is more, television programmers and advertisers are well aware of this. The reconciliation of these contradictory conditions—television's snubbing of minority elders and their persistence as audience members—underscores the place of nondominant elders as citizens who are multiply positioned at society's margins. Even if they are males, they are past the age of social dominance—youth and middle age—and they belong to an ethnic or perhaps sexual group that is marked as something less than correspondent to society's dominant members. What is more, they came of age during periods when belonging to a minority—especially one defined by sexuality—

meant submerging or perhaps masking one's social identity.[5] Nonetheless, the lack of popular media attention paid to people who occupy subaltern positions does not suggest that they lack strategies for using television to their advantage and that they do not formulate ideas about television that are important to them. This chapter tells the stories of thirteen elders who have developed their own understandings from their diverse positions as members of minority groups.[6]

Minority Representations: Appropriations and Responses

The television representations that members of minority groups have faced and the coping strategies they have relied on as a result are anchored in the ignoble history of the West's overwhelming tendency to marginalize nondominant people in its popular culture. Responses of those so marginalized have come forth faithfully, although objecting voices have been comparatively faint when measured against the blare of mainstream representations. For example, white American mythic images of African Americans, perhaps the most notorious of this society's misrepresentations of a subaltern group, did not, of course, begin with U.S. slavery. The practice can be traced to an older set of European ideas (and further back yet, if we were to be exhaustive). The European colonizing of Africa produced ugly stereotypes that have circulated widely in the last two centuries. Eventually, during the mid-twentieth century in South Africa, popular stereotypes inserted brazenly into such outlets as police information brochures polarized the Africans simply as either domesticated subjects (good) or terrorists (bad). Absent from image-driven products such as postcards and billboards was any mention of the portion of the South African population that had provided the labor to construct the hotels and roadways expected to draw in the white outsiders. In an especially abusive stereotype, European political humor has commonly depicted Africans as cannibals without any parallel effort to understand this practice. African intellectuals and literati have responded in mounting volume by circulating narratives about the conquering "barbarians."[7]

The Aboriginal-European situation in Australia and its environs has produced a similar set of circumstances. As Australian literary scholar Lyn McCredden has put it, the Anglo colonizers took Australia as a "blank page" and inscribed it with flat summaries about its indigenous occupants.[8] Postcolonial Aboriginal writers, on the whole, have found Eurocentric publishers skeptical of the aesthetic worth of their responsive poetry and fiction. Writers who have managed to beat past the impediments against broad-based distribution of their work have questioned dominant ideas about Australia and have supplied redefinitions of marginality by rejecting the authority of white critics, McCredden asserts. In the United States, a recent reader entitled *Indigenous Australian Voices* collects poetry and prose by Aboriginal women and men. This volume, through the diversity of its authors' positions, counters proselytizings about Aboriginals as a simple, homogeneous group and answers the practice of much trendy multiculturalist literature to ignore Aboriginal voices.[9]

Responses to colonialist appropriations of culture are easy to spot in art and literature, but what about in the everyday lives of ordinary people? For most people, participation in the production of commercial cultural products is an impossibility, but people do, through acts of cultural consumption, wage some degree of productive participation. As Kobena Mercer has suggested, for example, Afro-American (and Afro-British) hairstyle trends have been an instrument for commenting on the social politics of race. The Afro itself was a means of calling attention to the "Black is beautiful" ideal and rejecting the preferred model of black submissiveness, Mercer notes.[10]

The appropriation of images by a dominant culture may result not merely in stereotypes but also in co-opted meanings that get massaged to serve the status quo. Such neutralizing assumption of meanings can be easily recognized in the appropriation of indigenous music for commercial consumption. I recall in the late 1980s that the supermarket megachain Publix, based in Florida, ran a television campaign with a theme jingle incongruently set to a Muzak-style reggae tune. It was white reggae, devoid of the form's political charge and mobilized instead to push melons and meat. Capital-inspired thefts of symbols from marginalized cultures carry limits, however. Personal encounters between white youths, who have tried to appropriate the signs and symbols of hiphop, and Rastafarian subcultures have sometimes met vociferous objection from members of those subcultures.[11] Reactions to instances of appropriation have included the persistence of such discourses as those found in black rap, discourses that emanate from oppressed positions and which continue to criticize social inequalities. In fact, even as rap persists in its transmittal of oppositional cultural values, a predominantly white audience at once consumes and blurs its meanings.[12] Ridiculous, coopted, and ugly images of African Americans continue to occupy the texts of American popular culture, but African Americans have found means, such as the production and distribution of rap, to resist the hierarchically transmitted images of themselves that they see in the more established popular media. More commonly, however, African Americans have been faced with the practical need simply to resist commercially styled images of themselves—to take such sign systems as the Stepin Fetchits and the Perennially Second Banana of the Buddy Picture (hardly ever the leading man) with a grain of salt.

People belonging to minority groups, then, can find their status as members of the television viewing audience to be filled with tacit challenges as they dovetail what is often a favorite medium into their lives in practical ways. As Roger Silverstone has theorized, people construct their own space for television as they lodge it into personalized patterns of ritualized, rational consumption. In constructing this fit, they select tactics that serve their "nomadic" positions as viewers and, more generally, as complicated social beings. A person whose identity is shaped by an interplay of tensions—between pulls of color, gender, class, sexuality, religion, and perhaps disability, for example—may find herself or himself impelled to "travel" as a viewer to navigate meaning amid those overlapping, shifting, and frequently competing cultural categories. Janice Radway, in developing the ideal of the "nomadic" subject, acknowledged that people's positions are never fixed with respect to texts, because not only are the people and their circumstances changing, but differing aspects of audienceship are made salient for them at different times. Just as the postmodern condition has guaranteed for us the instability of a viewer's subject position, these competing markers of social identity tug viewers first in one direction, then in another. Viewer responses to the Clarence Thomas hearings and the O.J. Simpson trial were typical of this condition, with people's positions being tugged by their various social identifications. The plain stuff of everyday living involves people charting practical responses to the discursive tensions they are feeling. This means taking up perspectives that always represent an uneven laboring against the pressures one feels to respond to culture—to live in culture. Again, as Silverstone has admonished us, theorizing the wobble of everyday life does not call for romanticizing a "free" audience.[13] Identities may be unsteady, but they are constrained. Social class, for example, may be both an unclear, mutable state and, as British cultural scholar David Morley theorized in his classic *Nationwide* study, an undependable indicator of audience interpretations, but social class *does* bear on reception, as do multifarious other influences.[14] One of the most significant resources for reception is marginality. As I have suggested, old age combined with other oppressed positions works to exaggerate marginality, and this is where it becomes difficult to talk generally and routinely of people's interactions with mass culture.

It is inconvenient to theorize people as belonging to more than one nondominant category. When we speak of people as others, a simple label—old or female or black or poor—may strike us as having the greatest utility to get across the message of what life must be like to be that other. While it is difficult to represent people in complex terms, such representation is necessary to get close to the real human condition.[15] Just as scholars such as Janice Radway and Ien Ang have demonstrated the artificiality of labeling people as audience members, we are challenged to understand the so-called elderly audience as more diverse and complicated than mechanistic labeling would allow.[16] As British cultural scholars Jenny Hockey and Allison James have noted, the very personhood of elderly people varies extensively, because individuals "occupy either/or social positions with respect to gender, race, age and indeed social class."[17] This chapter does not attempt to construct all the possible concoctions of categories that might conjoin with the category "elder." It does, however, convey something about the texture of old age in America—that it is beyond the sorts of simple media representations we have come to understand as normal, middle class, and white.

Among the Milwaukeeans I interviewed were members of two immigrant groups: Laotian Hmongs, who had come to the country about a decade earlier, and Russian Jews, who had been in the area for about twenty years. Of the immigrants, only the Russians spoke functional English. Other people interviewed were Native American men and women, two gay men (one white, one multiracial), and two lesbians (both white). Among all these people, I found variations on several themes: a practical approach to television that was inflected by one's cultural position, a preference for seeing television as a social learning tool, and the tendency to champion images that resonated with one's own view. However, as should be expected, each group's views bore the markings of a special set of cultural circumstances, as their stories here show. At the end of the chapter, I return to these themes and offer thoughts on the groups of elders as a whole.

Before proceeding with a discussion of the interviews, I wish to say some things about the contexts in which elders of social and ethnic minorities find themselves in modern life in the United States. Since World War II, the number of American elders who live alone has risen dramatically, with family changes in structure and in size bearing a heavy influence. This shift has hit older members of minority groups significantly as traditions of extended family support have diminished. After enduring lifelong economic hardships, members of minority groups, on the whole, face tougher living conditions than non-Hispanic white elders. At the close of the twentieth century, conservative government initiatives have begun to chew away the economic tether grasped by poor elders, whose lack of clout has figured them an easy target. Meanwhile, the accelerated pace of social change—from shifting models of housing and caregiving to radical developments in the health care system—has left elders who live in poverty in need of strategies to cope.[18]

One of the sites where coping needs are especially salient is in the fast-growing need to manage the use of communication technologies. For example, many of the oldest old—85 and over—have expected to live out their entire lives within a short distance of their birthplaces. They have witnessed a remarkable diffusion of their children and grandchildren into the national landscape, and they have radically undergone an acquaintance with long-distance telephone communication as a practice embedded in everyday life. For most of these people, calling long-distance was akin to sending a telegram: it implied urgency, because it involved an expensive and mystifying infrastructure. Many of these once long-distance-shy elders now think little of picking up their cordless phones and punching in a 1 before dialing the area code of a family member who has moved. If the elder has been graced with financial success, perhaps it is she or he who has moved to a new area code—in the Sun Belt. Even among elders who live in poverty, telephone communication has become a way to collapse the space between oneself and loved ones. These elders, from what I have been told by the people I have met, still see long-distance—like cable television—as a luxury but have lost their fear of indulgence in it. Some of them live with cataracts or without air-conditioning but still budget a few minutes a month to call Junior in Atlanta.

Like most elders, those belonging to minority groups have struggled to adapt to changes in communication technologies and forms. I was struck by the tendency of the minority elders I met to respond to representations of their groups—and the lack of such representation—on television. Like the South African poets who responded to European colonialists' stereotypes and the American black rappers' insistence on maintaining the political charge of their music against the purloining tendencies of whites, the elders whose stories are related in this chapter recognized that their positions were marginalized both in the broader social context and within the field of popular representation. Like the poets and rappers, they had given thought to strategies for coping with these matters. Most often, such coping meant a refusal to take images of oneself—and of other marginalized groups—at face value.

Gay and Lesbian Elders

The gay elders I met were activists to one degree or another, belonging to a group called Sage, Senior Activists in a Gay Environment. As a result, most enjoyed connections with substantial networks of gay friends and led active social lives. They differed from the sort of gay person they all told me about: the closeted elder who, fearing reprisal from authority and community, kept his or her sexuality a secret in order to protect against real or imagined discrimination. As a result, I want to be clear that the stories I'm relating here do not sum up in any general way the experience of gay elderly Americans; they only show how such an experience might be flavored. Several of the elders I talked with made this point carefully when they talked about their own lack of interest in television in comparison with that of some of their contemporaries.

As Byron, one of the gay men I met, put it, he could "take or leave" television, but he knew that the medium held critical importance for some gay elders:

> I think generally, the older the gay person, the more in the closet they are. This is because they grew up at a time when gay people were viewed as sick by the medical and psychiatric professions. They were considered criminals by the police and the legal profession. And they were considered sinners—unspeakable people—by the religious community. And they were so completely ostracized—you could lose your job, you could lose

> your family, you could lose everything if anyone ever found out that you were gay. So everybody led these deep lives in the closet.
>
> Many older gay people are very much penned in. When you are an older gay person, and you try to do traditional activities for elderly people, it often doesn't work out. You go to a senior citizen center, where everyone sits around and talks about their grandchildren, or, if you're a gay man, you're looked upon as a good catch, a prime candidate for marriage. In these settings, it is difficult for people to disclose that they are gay. So lots of older gay people are isolated at home. Many of them would be absolutely, totally lost without the television. It's a companion to them, it's on all the time, and they don't know what they'd do without it.

Byron theorized that very elderly gays, on the whole, may be much more dependent on television than heterosexual elders, who also use the appliance frequently for company.[19] Elder gays, he pointed out, often suffer from "absolute, terrible loneliness" because their life partners have died, they have no communal channel for their grief, and they become socially marooned. Age-related impairments underscore their tendency to become homebound. Byron related a story about a woman he knew, explaining that many older gay people feel entirely disconnected from the gay movement:

> This older lesbian was being seen by a social worker, and this social worker suspected that she was a lesbian but didn't know how to approach it. Finally, during some very gentle prodding, the thing came out. This woman had had a lover who had died, and for six months, whenever the social worker was going to come, she ran around and took all the pictures of her lover down and put them away. She was so frightened that this woman would find out, and she would lose her benefits.

So, as Byron noted, sometimes very old gay people feel intimidated by authority even if their immediate fears are imagined, because they have lived through times when revealing their sexuality was dangerous business, and even among people with whom they spent a lot of time it seemed risky. That there might have been laws protecting her against losing her social benefits did not assuage this woman's insecurity.

Ned, a 66-year-old gay man, felt freer than the elderly woman in Byron's anecdote to reveal his sexuality, but only very selectively. Many years following the death of his longtime lover, Ned had become more detached from the gay social scene and remained close to only a few intimate friends; in addition to citing the end of his sex life, Ned also said he sensed an ageism among younger gays that made him feel unwelcome in settings dominated by them.[20] As a recovering alcoholic with many years of sobriety, he spent much of his time alone in the tiny, dim-lit, shabby office where he worked or in his studio apartment. He looked to the television for entertainment and companionship most evenings. As he approached old age, Ned noticed that his interest in television programs had changed little. For more than twenty years, he remained loyal to a single program—*Star Trek* and its various incarnations—but followed other dramas and comedies from time to time.

"The only two people in my personal life who know I'm gay are my doctor and my minister," he said. "But I never make up girlfriends," he added. Ned attended meetings of Senior Activists in a Gay Environment, and he had had other homosexual friends over the thirty years or so that he had lived in Milwaukee.

He felt comfortable partitioning off the areas of his life where he felt uncomfortable claiming gay identity. For example, Ned was a Korean War veteran. He was also opposed to the push for legitimizing gays in the military, and he adamantly opposed allowing gays on sea duty ("because those who feared them had nowhere to run"). He had had two long-term relationships with other men, but he opposed legalizing gay marriages ("because they forced people to deal with it"). He felt that he must keep his homosexuality secret in church, but he bragged to fellow parishioners (many of them older women) that he was a practicing nudist at home. In his dingy office, I suspect that he would have felt embarrassed to receive visitors from his church; there, alongside framed personal photographs, including a portrait of him in military dress from the 1950s, hung sexually provocative posters. Among these were a poster from the controversial 1996 Calvin Klein campaign depicting a youthful male model partially undressed and a large poster depicting semen splattered on the backside of a young man, the image cropped at the head and buttocks. The porn, as Ned explained it, was art, was enjoyable, yet at a different point in our interview, he complained that much television content was trashy sex and violence, and he refused to watch it.

Along the same puzzling line, Ned, whose ancestry was African, European, and Native American, was an acknowledged bigot. He identified with Archie

Bunker, not George Jefferson, from entertainment television, and he mistrusted blacks in real life because he said he feared that many of them were drug pushers and thieves.

In addition to running his small business, whose profit margin was narrow, Ned managed the apartment house where he had lived alone for many years. In his own studio apartment, he kept an old television set, and he had to get up from his recliner and go over in order to turn channels. He read the metropolitan newspaper but disparaged its quality and frequently bought the *New York Times* to keep up with business and science news. He also subscribed to a few periodicals, mostly business-related. Ned considered television a poor source for news—its reports amateurish—and he delighted in getting his serious information from print media. The television, he said, entertained him, and little else, although he said he liked learning from documentaries on public television about no particular favorite subject. Despite a penchant for science fiction TV, Ned showcased a middle-brow bias against what he considered television's biggest taboos: superficial news coverage, annoying commercial interruptions, and "stupid" content:

> I hate television news with a passion! . . . They will give you a, well, there's a comedy routine that I think summed it up: "The world is going to end—Details at 10." That's about the end of the story, that's it, just sound bites. CNN, I understand, is a lot better at it. I had cable at one time and dumped it—so many commercials. NPR is another one I like to listen to, not as often as I should, but I do like to listen. They will take a story and spend ten minutes with it.
>
> You can't tell a story in one minute. An airplane crash into someone's house—I caught the story. They spent two minutes. I mean, come on—two minutes—sixty people just died! How did they die? Well, it's a possibility that they got blown out of the sky by a rocket, and they just spend two minutes on the story and move on to something else.

Ned, like so many Americans struggling with a desire to keep what they perceive as TV's idiocy from escaping the box and contaminating them, faulted the medium's stupidity while acknowledging its allure. Many times, Ned found himself watching comedies geared toward black audiences. Although he expressed a passing familiarity with most of the black situation comedies, Ned was reluctant to say he watched them purposely: "Sometimes, I will turn the television on about ten minutes before *Star Trek* and it will be something like *The Fresh Prince of Bel-Air*, which I find an absolutely stupid show. In fact, I don't like any of the black comedies at all. To me, they're just dumb."

Although Ned could not say what he found "dumb" about black comedies, he finally noted that his own prejudice against African Americans probably motivated him to feel that way. A high school dropout who had received the General Equivalency Diploma through the military, Ned hated what he found to be an almost unavoidable self-identification with American blacks, who for him represented the underclass: "I consider myself middle class more psychologically than financially. My values are middle class but my income isn't." Almost in self-defense, he read the *New York Times* and watched PBS. "I've never stopped being educated," he explained, adding that he did not like to read books but found television an easier way to learn.

I asked him why he thought the quality of *Star Trek* was inherently better than that of the black comedies he disparaged. For one thing, he said, "I would like to go out in space." He added: "Science fiction is just a whole different world. Much of what has happened in the world science fiction has predicted years earlier. The whole era of the computer was written about in the thirties. The first computer was not created until about 1942." As he saw it, Ned could find a worthwhile escape in *Star Trek*, which was far removed from the possibilities of daily life, but not in *Family Matters*, which showed a black family succeeding against everyday, middle-class challenges. He found fantasy preferable to idealized portraits of the family—black families and white. (Ned disliked most other situation comedies as well for being unrealistic and lame.)

Science fiction was clearly Ned's favorite genre, but he preferred it mostly on television, in the form of *Star Trek*, *Dr. Who*, and other shows, rather than in book form, which he said he found too complicated to follow, especially as an older person. He ventured out to see most of the *Star Trek* movies, usually by himself, and said he had no friends who shared his interest in the genre, so he kept his thoughts about it mostly to himself. In addition to occasionally going to the cinema, Ned liked to rent tapes to watch on his VCR; he preferred male-action movies. "My favorite is the type where at the end of the first fifteen minutes at least two hundred and fifty people are dead by some gory method like blowing up an airplane. Personally, I'm not that kind of person, but I'm just fascinated by the idea." He preferred to rent tapes rather than go to the cinema because he liked to "sprawl out on the couch."

Although Ned said he watched few nonaction films, he had seen some gay-issue movies, such as *Jeffrey* and *Philadelphia*, the latter of which he called "one of the best." But television films about gay life, said Ned, are more powerful than cinematic stories. He mentioned *My Friend Andre* as a "realistic" version of everyday life for homosexuals, although his comments bore evidence of his concern with gay life as a male issue:

> The gay characters are not flighty—trying to make like they're women. Most gay people do not have any illusions about being a boy or a girl and have no desire to be the other. In older days if there was a gay character in a movie it was always a screaming queen—pardon the expression.[21] But the one complaint I have is that these movies always seem to be about very wealthy people. The majority of the gay community is not wealthy. But they show these wealthy people. And television is guilty of that, too. A lot of the scandalous types of behavior that constitute miniseries involve wealthy people.

I asked Ned why he thought that was. He answered: "I think there's this mistaken belief that lower-class people simply don't have this sort of thing occurring in their lives. Or maybe it's because by making it wealthy people, you're not insulting the audience—you're saying that only wealthy people are decadent enough to do this sort of stuff."

Ned said he would prefer more depictions of ordinary people on television, gay or otherwise. He particularly liked Estelle Getty's character on *The Golden Girls*, Sophia, the aged, plainspoken immigrant Italian woman who practiced stereotypical "old-lady" behaviors such as carrying her purse with her from room to room in her own home. "She was a bitch," Ned explained. "I don't know that there were many nice things about her character. . . . But for me, all women represent authority figures. Older women scare the hell out of me."

Ned wasn't sure why this was but said he was raised by an adoptive mother, also multiracial, who, he said, was very gentle and the furthest thing from autocratic. Males, he said, had been entirely absent from his upbringing, so he found elderly women figures such as Sophia more or less realistic. Ned ambivalently longed for "realism" while rejecting portrayals that made him squirm, badly wanting to "read himself" as a middle-class white gay elder.

I met Mildred in the small east-side Milwaukee apartment where she had lived for only a few years. It was hot, and she had opened the windows to the noisy city street, lacking air-conditioning. From the looks of her home, it appeared that Mildred had spent the majority of her disposable income as a feminist academic on books, which she enjoyed for pleasure and for pursuit of her research in urban studies. She told me she could not afford cable for her television sets, of which there were two. She watched them in total only a couple of hours a week and had acquired them as inheritances from her mother.

Mildred had grown up in a middle-class, white family of three children in the liberal, progressive state capital, Madison. Her parents had resisted buying a television until she was in high school in the mid-1950s. Even then, they rarely watched it, and Mildred had watched little television in her life. She went on to marry and raise a family before divorcing her husband and, as she said, "coming out to myself" after realizing that she had always been interested in other women, through her participation in feminist organization efforts, as contemporaries and as the subjects of her feminist theoretical interests. Her public life as a lesbian was limited to professional and social circles; one of her children, her son, did not know, she said, although her daughters did.

Mildred had a wide circle of feminist and lesbian friends, and she served on the board of the local senior gay activist organization. She liked going out, especially to readings among other women at the local feminist bookstore and to films, although she often attended the latter alone. She went out several evenings a week and spent much of her other evening time reading at home—in the living room, in the bath, and at bedtime, to get to sleep. In addition to trying to conquer her mounting book collection, Mildred made daily stabs at her periodical subscription pile. She also listened to National Public Radio, preferring it over the local newspaper, whose reports she found both pedestrian and depressing: "I like keeping an ear out for what's going on while I'm doing something else, and then if it is something I really want to hear, I can turn my attention to it."

Unlike most women who have been the subjects of media audience studies—including some of the single women interviewed for this book—

Mildred made no excuses for granting herself the time she wanted to watch television: "Not in recent years, anyway. I'm at that point in my life where I really don't have those kinds of domestic demands. I have no problem spending time doing what I want to do."[22] Having doubly shed her role as heterosexual wife, Mildred could watch *NYPD Blue* without simultaneously folding laundry to justify—for herself or for anyone else—the expenditure of her time. She felt very comfortable focusing on television. She could talk about particular scenes and characters in vivid detail. In talking of *NYPD Blue*, for example, Mildred talked at length about character development over the life of the series as well as the moodiness of some of the street scenes. Such a tendency was no doubt fed by her own exposure to the humanities in academia and may have been informed by her insider understanding of the research process. (As much as an academic might like to serve as a "good subject," it is impossible to submerge ideas of the kinds of "data" a researcher might be looking for.)

Even if she took a certain interest in the medium, television did not seem terribly important to Mildred. She did like to pick out one or two shows that she had seen favorable criticism of and "follow them" for a year or so. Some contributing factor, such as seeing the New York scenery in the case of *NYPD Blue*, the main character's feminist development in *Roseanne*, and what she deemed the clever dialogue in *Coach*, generally also drew her to a particular show. But she refused to become absorbed in more than an hour or so per week, because she found it wasteful of her time and less interesting than the many other activities she could engage in.[23] She often would set her VCR to tape her show if she made plans to go out when it was on. And she checked the Sunday newspaper to research what other programs she might like to see: a documentary on PBS, a special show featuring a gay character, and the like. She found episodic series more workable than a serial: "I'm just not that committed to watching television. Once in awhile, I'll watch *Masterpiece Theatre*, although not very often because I'll get into one and then I'll miss several of them, and I'm not a night person who will stay up until midnight to catch up. I do like *Prime Suspect*. I'll go out of my way to watch those, even stay up." Jane Tennyson, the renowned feminist police detective of PBS's *Prime Suspect*, struck a chord with Mildred, because so few characters like her were to be found on television. Mildred enjoyed reading feminist fiction, and her affection for the *Prime Suspect* stories matched this interest.

Although Mildred's contact with television was much more limited than that of most of the elders I met, she did mention favorable impressions of a number of characters when I asked her about television's portrayal of the elderly. Not surprisingly, all the characters were women. I asked her about the character Ruth Anne, the wise septuagenarian from one of the shows Mildred had followed in previous years, *Northern Exposure*: "She's a good character. She was an older woman who was presented as kind of eccentric, which a lot of us would like to be in our later years, but not someone who needed to be certified or institutionalized, either." Mildred admired Ruth Anne's autonomy, and she liked television's capacity for representing such a character nonstereotypically.

On the other hand, she found television's portrayal of gay people and gay-related issues almost insulting. "It's trendy," she explained. "Every movie has to have a gay or lesbian person in it. You can plug it into a show and say, 'Aren't we progressive?' "

But she did appreciate television's exposure to gay concerns: "*NYPD Blue*, as you might know, has an ongoing gay man [character]. I think they do a pretty good job the way he is presented." She went on to say that the character, who is portrayed as effeminate and officious in his role as the office secretary, might come across as a stereotype, but she praised the show for having an episode that focused on this caricatured image and problematized it.

Mildred had heard or read about several other "trendy" moments on television featuring gays or lesbians but had made no special effort to see them if they seemed overly commercialized and her calendar offered other commitments. Roseanne's highly publicized gay-bar kiss seemed an obvious ratings ploy and did not hold sufficient lure to keep Mildred home. She felt the medium could comment on and capture pictures of real life, but, to be sure, it was not *the same as* real life.

While Mildred's commitment to television was on limited terms, she said she would not want to be without it now that she had accommodated it into her life. "If mine broke, I probably would buy another one, just to keep following my show, whatever it might be. Or there might be something on that I want to see." Like most of the single elders I met, television viewing tended to be a solitary experience for Mildred. She had had a relationship for a few years with a woman who was a much bigger TV fan than she was, and their differences caused a problem. "She wanted to watch pretty much all the time," Mildred said, "and I wasn't crazy about it. It got old."

Ned and Mildred may have partially hidden their homosexuality from people they knew in everyday life, but Cynthia was more open, keeping her guard on for official circumstances. Her openness resonated with my own understandings of lesbianism dating from my growing up in the working-class South, where it seemed much more common for women to be out about their homosexuality than it was for men, for whom there may have been harsher social penalties. Cynthia, who had grown up in a small town in Tennessee and spent her teen years on the East Coast with her grandparents, married at 16 to a boy she barely knew because "it was the thing you did." By the time she discovered her sexuality, "it was too late." She already had a son. As a young wife, she moved with her husband to small-town Wisconsin to be near his relatives and before long learned that the husband was having an affair with his high school sweetheart. Cynthia left, leaving her son in the care of her grandparents. Failing to find employment, she joined the army and stayed to earn a pension. Had her grandparents been able to continue to care for her son, she would have stayed on indefinitely; she liked travel and enjoyed the routine and organized sports. But when she had to quit the army to finish raising her son, she moved to Milwaukee and found factory work. Finally, the opportunity presented itself for her to earn a licensed practical nursing certificate, and she worked in hospitals subsequently. During the majority of her nursing years, she and another woman lived together before the woman died of cancer. Cynthia had a few failed relationships afterward, one with a much younger woman, and then settled into a lesbian friendship with a woman about her age, although they lived apart. The two often spent time together and also socialized with the woman's former lover, an older woman who at that point lived in a nursing home. Semiretired in her mid-60s, Cynthia also worked part-time as a bartender at a local gay bar.

Since entering her elder years, Cynthia had become a gay activist, having, among other activities, appeared as a model for a cover story on aging gays in a local alternative magazine. Cynthia felt savvy about "the system." She trod cautiously where her army pension might be threatened but was active in state and local political issues, mostly as they related to concerns of elders, especially those who are gay. Cynthia's self-confidence and liveliness sustained her self-reflexive, soft-spoken statements. Her style was underscored by her appearance: she was small but sturdy and participated fully in butch fashion. She joked in our interview about the irony of having to put on face makeup for her magazine photo shoot.

Of the four gay elders I met, Cynthia was by far the biggest TV fan. She had had a VCR since the mid-1980s, had long subscribed to cable television, and depended religiously on two cable viewing guides. She expressed enthusiasm about many types of programs. For one, she was a mystery enthusiast. She looked forward to the *Sherlock Holmes* and *Miss Marple* programs on PBS and A&E and had been such a devoted *Murder, She Wrote* follower that she taped the show if she had to be out when it was on, whether it was a rerun or not. Although she enjoyed seeing older guest actors in the show ("because it was nice that they could get work"), Cynthia especially liked seeing the young guest stars, because "Got to keep contact with the young to stay young!" She talked about her mystery-viewing experience: "If the phone rang, I'd say, 'Damn!' and sometimes I wouldn't answer. I have voice mail—they can leave me a message." Cynthia often watched mysteries with her lover, who also liked the genre, but she mostly watched television alone in her small house.[24]

She was a self-proclaimed movie buff and said she especially enjoyed "anything with animals or Indian culture, because I'm a quarter Iroquois myself, and I enjoy learning about it." She also liked documentaries for the same reason. Cynthia had not known her Iroquois grandfather, and soaking up her heritage through television was both convenient and rewarding for her. "I take the bad portrayals of Indians with a pinch of salt, I guess," said. "It's no different from everything else."

She also liked her cable premium channels. "I'll take one for a while, then switch to another if I think I've seen all the movies. I think it's a shame that a lot of elderly people can't afford cable, but I'm going to take it as long as I can afford it." Cynthia felt unbothered by explicit sexual and violent content on television "as long as it's really a part of the story." The quality of the story and whether its outcome made sense were more important than not being able to identify with a main character in a fairy-tale heterosexual romance.

Cynthia said she spent many hours watching evening game shows and daytime talk shows, depending on her work schedule. Although she found *Jeopardy* too hard, she liked solving the puzzles in *Wheel of Fortune*, because they made her feel "pretty good for my age." She had stopped watching programs with male hosts, whom she said did not interest her, but she was quite fond of Oprah Winfrey, Ricki Lake, and Rosie O'Donnell. "Rosie is gay, so they say," she noted with a grin. Rather than hold a grudge against that system, Cynthia reveled in incremental victories.

Cynthia enthused that the media generally and television specifically have begun to treat gays "more respectably." She cited feature coverage in the local newspaper of Milwaukee's gay Pridefest event, although she added that elder gays rarely attract the attention of the media. "I like the show *Friends*, because it shows an intermingling, at least, between gay and straight people," she said. "If we can see more intermingling, that would be a positive step for television, a positive step for society."

Cynthia said she had received information about gay celebrities for many years from "underground papers, smut papers" she had read off and on. "In the gay community, there was an awareness that Rock Hudson was gay. I'd think when he would be cast as a leading man, 'Too bad you don't have a boyfriend there instead of a leading lady!' " Her comment came with an impish smile. Cynthia wished that television would do a more exhaustive, more accurate, more generous treatment of gays and the elderly, she said, but she did not get upset about it.

In addition to her home viewing, Cynthia said she watched many hours of sports each week at the gay bar where she worked. "Packermania is a big thing for us at the bar," she said. "Everybody wants to see the football game." At the bar, rooting on the home team was a cause shared by the bartenders and customers regardless of their gender or age, a social event in which they were able to find unity and glee, Cynthia observed.

Cynthia saw television as an instrument for enjoyment as well as a player on the field of social change. She hesitated to take its power extremely seriously but cheered what she saw as its growing progressiveness. The mixing of gay and straight social relationships in television fiction and the entry of a perhaps gay talk show host through the back door were signals to society that gayness was normal.

Byron was the most "out" of the gay elders I interviewed and was, coincidentally, the least enamored of television. In his mid-60s, he continued to work as a stockbroker with some success, living in a turn-of-the-century east-side Milwaukee home. The house was in some need of repair, but it was stocked with antiques and an extensive collection of colorful Victorian glassware and artwork. Byron lived alone but had a huge network of friends, mostly other gay men. He had ceased to be sexually active and had never maintained a longtime monogamous relationship but had accumulated a sprawling gay family of whom he was quite fond. In addition, Byron remained close to his sister's family, relishing the role of "Uncle Byron." But he had put much emotional distance between himself and his small-town Indiana upbringing. He discovered his sexuality as a teen and found provincial small-mindedness difficult to bear. A quick solution was to migrate to Gary to work briefly in the steel mills shortly before he was to have graduated from high school. Then he moved to Milwaukee and never looked back, glad to have found the emotional security that a big city could afford him. Byron was open about his sexuality to his bosses and his clients, who, he said, never questioned it, because he made them money and that is what they paid him for. He was an affable, plainspoken man who had channeled much energy into gaining public recognition of gay elders. In this effort, he had founded the local chapter of Senior Activists in a Gay Environment and had appeared from time to time on public-access television shows about gay elders.

Byron largely avoided television because he found its content to be "trash." He thought the portrayal of sexual relationships on prime-time television was inappropriate as popular fare. This was quite apart from interests he may have had in popular culture outside of network television. At the same time, Byron said he wished that Hollywood film and television would more realistically treat gay people's interests. He mentioned the popular film *The Bird Cage* as an example. "It was just dreadful. They made the characters into stereotypes." Like the other gay elders I met, Byron liked going to the cinema to see films that depicted gay characters. He said there were too few of these. He noted that he was tired of the conservative backlash against outfits such as Disney for trying to recognize gay individuals as regular people. I asked him if he thought Hollywood should reframe the clichéd heterosexual romantic story more often to tell stories on film and television about gay couples. Did, for instance, the classic boy-meets-girl Disney text rankle him? He put the issue in perspective this way:

> I'm not sure I would be interested in a gay Mickey Mouse! I don't think that it's necessary—I just don't think that it's necessary. This is something that many people mistake—most gay people feel the way that I do, that is

> that heterosexuality is wonderful—we don't see that there's anything wrong with it. But it just doesn't work for us. I would like to see a gay character here and there that was brought into the thing in a very natural way, the same way it happens in real life. We don't want to be exploited or forced into a situation, but just see a natural gay character some of the time. I don't want to rewrite the Hollywood romance, to make Doris Day a man.

The kind of "natural" stories Byron said he would like to see was the introduction of gay characters into ensemble casts such as *Northern Exposure*, the long-running CBS drama series about fictional Cicely, Alaska: "I thought the treatments of gayness there were pretty sensitive. The two men who had the hotel or restaurant or something. I thought that was extremely well done, and I did watch those episodes because I heard about them and I wanted to see what they'd done with them." He also liked the elder character Ruth Anne, "a laid-back old lady who has seen it all, who can't be shocked." Thankfully, Byron said, his mother had been like Ruth Anne, and as his gayness came to light for her, she took it in stride. Some of her friends did not, and his mother, to her credit, cut loose those relationships.

But *Northern Exposure*'s gay story line or the appeal of Ruth Anne hadn't drawn Byron in sufficiently to make him watch week after week. He just wasn't that interested in television. He seemed quite knowledgeable—mostly from reading gay and popular publications—about which characters actors were gay, but he did not make an effort to watch them: "Oh, occasionally I will spend an evening—just throw an evening away—and I will channel surf and watch whatever is on. I will watch a movie occasionally."

In channel surfing, Byron noted, he frequently saw television, especially talk shows, exploiting gay issues, and he resented this. "It annoys me when they go way out. The talk shows have just about beaten the topic to death. Very frequently it's counterproductive when they sensationalize it." Echoing Ned, he added:

> It used to be that the gay person was always the one who did the murder. They were the social deviant. That's not happening so much anymore. Or the gay man was always the little flitty person and what have you, a stereotype. Now I think we are being a little more realistically represented. I think most people can see through the exploitation. Everybody knows that the reason Rodman carries a feather boa is to attract attention.

Byron liked seeing developments such as Ellen DeGeneres declaring her sexuality, noting that he wondered why Lily Tomlin had not done so. He understood from reading gay advocacy publications that Tomlin and her partner had worked to produce most of Tomlin's stand-up material for many years. But he did not favor "outing" celebrities or even politicians, unless those politicians had taken stands that were harmful to gays.

Popular media, Byron felt, bore a responsibility to do much better by citizens who lacked cultural clout, including gays and elders. The significance Byron laid at the feet of producers and programmers competed against his middle-class bias against associating himself with an often disparaged medium. He echoed Ned's own shame about TV viewing when he spoke of "throwing away" an occasional evening by indulging in rudderless channel surfing. This issue also arose in Mildred's discussion of placing stringent limits on her program consumption.

These gay elders were in many ways unlike the retirement community residents I met in connection with chapter 3 but voiced some similar opinions. These comparably younger gay elders lacked the huge time resources that seem to be available to many older people. They were all employed. This combination of work demands and the pressure of material or idealized class identity drew Byron, Mildred, and Ned to perceive television as an ambiguous combination of time waster and political instrument. The latter role was necessary for them to justify their own enjoyment of the medium, as it had for the Woodglenners. Significantly, Cynthia perceived television as having political power, too, but its politics troubled her much less than it did the others. Her working-class sensibility propelled her toward her "live and let live" philosophy. Television's representations of the cultural categories that these elders found salient—age, gender, sexuality—dominated their concerns, but the enormous shaping of their perspectives by class identities signals the ever mired nature of any sort of handy explanation of people's relationship to mass culture. The bleeding between the cultural categories that constitute us as individuals does not point merely at the impossible finality of the interpretive project but at its ongoing necessity, lest we begin to draw neat boxes around people such as Ned, Mildred, Cynthia, and Byron.

The four gay elders' experiences with television seemed to me richly distinct from one another. One reason for this may have been that they, like myself, were all born Americans, native English-speakers, and, despite their subaltern sexual positions, identified in so many ways with dominant culture and appropriated its symbols so facilely. I found it easy to communicate with them because our universe of knowledge and our interests were in many ways compatible. In contrast, my interviews and those of my research assistants with members of the other minority groups I sought to represent in this chapter produced transcripts whose themes clustered more within each group and produced less sprawling tales, on the whole, about the individuals themselves. Helpful translation assistance undoubtedly preempted development of the rich personal narrative.

Russian Jews

Milwaukee has long been a destination for immigrants from Russia and neighboring countries. Thousands of immigrants, most of them Jewish, moved there following the dissolution of the Soviet Union, but many had lived there for decades before. A large portion of the immigrants settled on the city's east side and in its north shore section. The Jewish Community Center is a huge complex on the east side serving thousands of the metropolitan area's Jewish population, including several hundred Russians. The center puts on a variety of activities for elders and provides bus transportation. It is one of the places where the Russian immigrants can meet, but an opportunity that many of them consider even more important is meeting non-Russians: English-speakers who have their roots in European Jewish culture but who represent their goal, American citizenship. Languages spoken between participants at the center include English, Hebrew, Yiddish, and Russian, among others. The center has provided a space for many elderly Russians to maintain a sense of their history while making a bridge to the present.

Among the Russian immigrants who were active at the center at the time of my research were Rachel and Ayzik, both of whom had lived in the Milwaukee area since the late 1970s. Ayzik, 78, came from a working-class family and had begun working at age 14 in Russia to help support his mother and three sisters after his father died. After coming to Milwaukee, he lived in a rooming house for Jewish elders near the center but later moved to a modern apartment in a western suburb with his Romanian caregiver. A major cheerleader for the American dream, Ayzik refused to watch Russian-language videos that some of his contemporaries liked because he feared that doing so might impair his English, which he had started learning at a local community college after immigrating. He was now an American citizen and taught English at the center, although he also enjoyed performing Russian Jewish and Yiddish folk songs there. He did not read local newspapers but did receive Jewish publications in English and Hebrew to keep up with the news; he was especially interested in events in Russia and Israel. Ayzik was proud of his tidy apartment, which was stocked with such modern conveniences as a twenty-seven-inch television set with a cable converter, housed in a wooden entertainment cabinet. He used a remote control and cable guides for the set. Ayzik also had a boom box CD player, home video camera, and microphone recorder, which he used to practice his songs. Along with such titles as *Tootsie*, *Kindergarten Cop*, and *Delta Force*, Ayzik had in his extensive video collection a five-part series on the Holocaust, which he had taped from network. He scattered touchstones of Russian culture, folk art, mostly, throughout the apartment along with photographs of his Russian American children.

Rachel, also in her 70s, was, like Ayzik, widowed after a long marriage. She spent the daytime at the center and spent most evenings alone in her east-side apartment, where she had lived alone for many years since her husband's death. Her children and grandchildren were scattered across the United States. She was a gregarious, vigorous woman who liked to stay at the heart of activities. She laughed easily, seemed to have many friends, and liked to participate in the shared knowledge of what was going on among her peer group. Her activities were, like those of many "involved elders," confined to the daytime. She liked to be home before dark because of her justified fears of crime in her neighborhood—she had been mugged there twice. Rachel found it impossible to read and use many forms of media, because she was legally blind. She had a large console television set in her living room and a smaller set in her bedroom.

These two Russian Jewish elders were big media users. Ayzik's use was broadly divided among television (including movie videotapes), newspapers, and musical sources, while Rachel was wholly involved with television. Both had had television sets in Russia, at different times. Rachel recalled having a small black-and-white set in the 1940s and watched mostly American and European movies, dubbed in Russian, but in later years she had no TV. Ayzik had had a television set for many years in Russia and enjoyed it but said he had not been prepared for the bounty of programming he would see in the

United States. Both acquired sets immediately after moving to Milwaukee and had employed them to help them learn English.

While Ayzik's language acquisition was supplemented by community college classes, Rachel depended almost entirely on American television to learn the language of her new country.[25] Her English had become quite good, but she remained devoted to television and left her set on most of the time. During the day, she set her VCR to record four hours of soap operas on ABC, the network favored by other immigrant center-goers, and listened to them in the early evenings and at night, splicing in her viewing of prime-time news, dramas, and comedies that might interest her. Reluctant to say she might sit alone and watch these programs without doing other things, Rachel said she generally watched while she occupied herself with "jobs" around the house.

The soaps held a special charm for her: "You know, it's real life, too. I like to observe life, you know. I want to know what will be later and what will be. I could tape even more. I could tape five, six programs, but I don't." Rachel developed solid likes and dislikes of the soap opera characters, recognizing them as good and bad people, and for her they were quite realistic. "Sometimes I don't like the characters. 'I'm glad they kill him!' I think to myself."

Rachel did not usually use her remote control to zip through commercials. In fact, she found the commercials as entertaining as some of the content. She nostalgically recalled the "Where's the Beef?" campaign that Wendy's had run a decade or so earlier. Although she did not mention Clara Peller's name, she remembered the woman's character with a chuckle. Rachel said commercials influenced her shopping tremendously, because they taught her about providers of the goods and services she felt she needed.

Ayzik did not express enthusiasm for the commercials, but he said that, other than making him nervous when he was trying to follow a story, he appreciated the networks' need to show them. Ayzik's and Rachel's embrace of the commercial television model resonated with their full embrace of American culture. Both clearly believed they were living the American dream and were unwilling or perhaps even unequipped to criticize their adopted country. As Ayzik put it, "I saw that everybody could come to America. You can make a living if you want to be an American, and if you want to live here you have to work not just for yourself but for the country, too." Rachel expressed not so much a moral commitment to capitalism but a compliance with its model for supplying success: "It's nice to see on TV how they dress, how they live," she said. "I come from a country that didn't have too much glamour. It's nice to see that show The Rich and Famous." *Lifestyles of the Rich and Famous* had helped make Rachel come to believe that literally anything was achievable in her new country.

Ayzik reported similar feelings. Each evening, he watched television in his living room with his caregiver, starting religiously with *America's Funniest Home Videos* reruns on weekdays, then perhaps whatever movie they might find on their basic cable schedule ("for relaxation") and ending with the local news broadcast before retiring for the night. Ayzik also said that during the daytime he often watched *CNN Headline News*, which he identified by channel number only, because he liked getting an international perspective. "In Russia we have just channel A and B—that's it," he said. "But here we have seventy-two channels. And during the day you can change and change and change. They've got everything you want to see. I used to see just what's going on in Russia, in my country and city. Now I can know everything that's going on all over the world." He attributed American television's range of choices and its "eye on the world" image to the palpable freedom he understood to be intrinsically present in his adopted country.

He often used his remote control to click from channel to channel, always stopping if he saw *The Lawrence Welk Show* in reruns. The show represented for him an America he never knew but longed for: "It was another time. I like the music, I like the songs, how the people are dressed—everything is nice to see. And Lawrence Welk is a very nice man." He had no idea that the patriarch of champagne music was dead.

Ayzik far preferred the world of Lawrence Welk—and that of *The Barbara Mandrell Show*, which he evidently found on The Nashville Network in syndication—to much of what he saw on prime-time television and what he heard on daytime TV, which he mostly avoided. About sex and violence, he noted:

> Who needs it? The children, they are watching the TV. Why do you show to them? The mothers and fathers are going to work, but the children stay home and they are smart. They are changing the channels to what they should not see. My grandson did that, and he was saying dirty words. It's not for the elderly, too, this kind of television. Everything, they don't have

> to show. I don't have to know that he kissed a girl. And the gays. Lesbians. What kind of life they have. They show us, and it's terrible. I don't understand it.

Rachel's health problems kept her awake at night, and her bedroom television set kept her company. She usually left CNN on to keep abreast of world news, especially in Eastern Europe and Israel. Ayzik, too, reported watching television news often on his small bedroom television because of sleep troubles. This is an event that many elderly people have reported to me during the course of my research. Many elders live with chronic pain, and it often seems most pronounced at night. This can be a time of terrible loneliness. Television newscasters, talk show hosts, and infomercial presenters can seem like comforting friends.

I have heard many young and middle-aged people complain that their elders are "so dependent on television that they will even watch it all night!" For most of us it is difficult to imagine that, although we might feel exhausted, pain might prevent us from sleeping more than a few hours each night. The "company" that television provides may seem a poor substitute for human contact, but elders such as Rachel sorely feel a need for it. It is more practical for her to look at a smiling John Davidson peddling in an infomercial than it is to pick up the telephone and call her children. In this respect, Rachel is no different from other people—young or old—who watch home shopping hosts or psychic friends to alleviate some perceived pain or isolation. In Russia, Rachel would have been living in an apartment with her children and spending much less time alone in her old age. In America, she relishes the luxury of her own small home and readily compensates for its drawbacks with an American solution.

Native Americans

I interviewed three Native American elders, all of whom lived in working-class neighborhoods of Milwaukee's multiethnic south side. Of these, only Wiley, 72, was much of a television fan, and he used it mostly for watching sports and local and national news shows, common practices for an elderly American man. Wiley grew up in Minnesota on the farm his father worked as a tenant. He dropped out of school after eighth grade to work on the farm but joined the army for a chance to improve his circumstances.

He served for most of the 1940s and 1950s and lost most of his everyday identification as a Dakota Indian as he worked toward becoming "regular" army. The army allowed Wiley to complete the General Equivalency Diploma requirements and gave him the opportunity to play in baseball and football leagues. When he left the army and relocated to Milwaukee with his new wife, he learned the heavy construction trade, where he earned his living until retirement. For more than thirty years, he and his wife, who still did factory work in her 60s, lived in the same rental house in a rundown neighborhood near the church to which Wiley walked for his weekday congregate meals with perhaps a hundred other Native Americans. It was at the meals site, in a large basement, cooled by big, noisy window fans, that I met him. He smiled often in our interview, but it was difficult for me to understand him because of the fans and because he covered his mouth when he spoke in an apparent effort to conceal the loss of perhaps half his teeth.

I also met Red Cloud, a Winnebago Indian, in the noisy church basement. She came for the weekday meal, too, but, unlike Wiley and most of the other Native Americans who filed in shortly before the noon meal service and left quickly afterward, Red Cloud came early to set up her quilting project. She loved to sew and, I think, relished the admiration of those around her. Red Cloud made the striking quilt tops on a home machine but, despite difficulty seeing, meticulously hand-quilted the layered spread. She took minute, even stitches, in patterns she had learned as a girl, on the brightly colored piece quilt. Except for when I came and engaged her in an interview, I saw that Red Cloud kept pretty much to herself in the meal room. During the time we talked, she spoke quietly and dispassionately. I took her tone as a signal of a long fatigue and perhaps depression. As the conversation unfolded, she explained to me that her parents, her husband, several of her ten children, and herself had all been alcoholics. Many of these, including a son, had died of the disease, and the rest were in recovery. Red Cloud lived in the small house she had rented for thirty years, having finally bought it from the landlord to avoid having to move. A daughter, son, and a few of her fifty-five grandchildren lived with her. At 66, Red Cloud despaired that, unlike her younger sisters, she did not yet have great-grandchildren. She loved to sew for her grandchildren and hoped that the new generation would come along soon.

The youngest Native American I interviewed was Frankie, 62, a member of the Prairie band of the Potawatomi tribe. She and her husband, also a tribal member, had grown up in small Wisconsin towns before moving to Milwaukee for work opportunities. She told me that they had enjoyed life in their multiethnic neighborhood, but she had sad memories of being the object of small-town bigotry, recalling that she had made many good friends whose parents severed ties with her once they had learned she was not white. Milwaukee was better, but still she saw prejudice.[26] They were planning to sell the bungalow they had owned for many years and move to a planned apartment building for elderly Indians. Frankie, who had amassed fifty college credits in state programs for Indians throughout her career as a teacher's aide, had retired recently but now worked in a public program for disabled Indian elders. She took care of a different elder each weekday, including, on Wednesday, her husband, who had been disabled for many years. Frankie and her husband had two grown children and several grandchildren, whom she doted on. Frankie's cheeriness and chattiness marked our interview.

The most striking development from my interviews with the Native American elders was a tendency to see television as unimportant to them personally and perhaps something to be greeted with suspicion. For Red Cloud, this theme came through most clearly. Because she was a recovering alcoholic and suffered from nerve problems, Red Cloud said she found television difficult to watch because it required a focus she did not have. This was interesting, considering Red Cloud's ability to focus on quilting, but that, for her, was a different sort of mental focus. Television and reading required her to follow a narrative, to absorb new information quickly and successively, and she found that entirely too taxing. Additionally, because she was so involved in the lives of her children, who were not television watchers, she had little time for it. Red Cloud, who practiced a blend of the Christian religion to which she had been converted and the Indian religion of her upbringing, said her children had abandoned many of their "white ways" and rejected much of contemporary popular culture. "They went back to their traditional ways, although they weren't raised that way," she explained. "They do the sweat lodge."

Other than news, the only programs that Red Cloud said she liked to hear about and watch were those having to do with Indians or the West, and these were important enough to her to have cable television installed: "Sometimes there are Westerns on in the afternoon, and I like to see them. Channel 3 [TNT]. Real old, real old Westerns. I love Westerns. I like the style of the picture. When I see Westerns, I see these old chairs and tables, these old houses. I like the scenery of the old West. The Westerns were outside mostly. Natural." Watching the genre allowed Red Cloud to connect with her cultural past, but she said she viewed the story lines of the movies skeptically and generally disregarded them. "It's too bad how they portray the Indians. It's not true."

I asked her if this stereotyping bothered her. "No," she said simply. "Because I know it's not true." She expressed no worry about effects on others.

In addition to watching the Westerns, Red Cloud said she also liked to find specials about the West or Indians. She followed local and national news broadcasts to keep up with current events and to learn what the weather was going to be.[27] She did not expect much happiness to come from watching the news, and little of import to her, in fact, but she did like to feel informed. Newspapers seemed an unjustifiable expense.

Wiley, like Red Cloud, said he watched television perhaps three hours a day but found it not very important to him. He said it was a time filler in his life after losing so many of his contemporaries, although he and his wife often watched together. If his television set broke down, he said, he was unsure whether he would replace it. As it was, he watched mostly sports and local news reports, although when he was younger, he, too, had liked Westerns. "I liked to watch them shoot 'em up in the old West," he explained. What about the Indians always being the bad guys, always losing, I wondered. "I didn't think too much about it." He had especially liked Tonto in *The Lone Ranger*, who represented an honorable Indian figure. After all, even if the traditional cowboy-versus-Indian narrative did not give Wiley pause, it was nice to see one's race portrayed heroically in this critical historical period.[28] Wiley later enjoyed country-western shows such as *Hee-Haw* and variety shows such as *The Carol Burnett Show*, because he enjoyed the music and the folksy characterizations of the actors. As with Ayzik regarding Lawrence Welk and Barbara Mandrell, he found their plainness and naïveté appealing. But television, for Wiley, always took second place to contact with the human and natural world. "I prefer when I can to go hunting with my sons," he said. "Or fishing." As an alternative, he often watched nature shows.

Frankie watched mystery shows, such as *Murder, She Wrote* and *The Commish*. "Those are roughly my two programs," she told me. "Now, if I could have that big living room TV in the bedroom, I would probably watch the old movies on the American Movie Classics, but I don't get the cable in the bedroom. I would feel bad sending Mister to the bedroom, because I like to let him watch his sports on the big TV." The nicer television in the living room also had the remote control. Frankie's husband could not move around much because of his disability and derived many hours of enjoyment each evening watching baseball or football on television, mostly on cable.

While Frankie devotedly followed her "two programs," she said she found much of television tiresome. "All it is, is kill, kill, kill. My shows, they got some good actors there. They got a story. But there's not all that killing, the same thing over and over, like *Perry Mason*, which my husband watches."

Although *Murder, She Wrote* and, to an extent, *The Commish* were highly formulaic, in that way no different from *Perry Mason*, for Frankie they were distinct. The sameness of most of television bothered her tremendously, as did what she found to be gratuitous sex and violence. She found the news, particularly, to drone on about the same dangerous or superficial themes: "I like my news. But they go on with things. I get tired of it. The O.J. Simpson trial. I got tired of it. You keep hearing the same old stuff and there's things happening right in our city that we should be learning about. All of a sudden the bombing in Atlanta. Now that has taken two weeks."[29] The sameness of romance novels, by contrast, appealed to Frankie. Ritual, as in Frankie's television viewing and novel reading, is salient in the lives of elders, as it is for all of us, but what varies is where we find ritual comfortable to consume. Some popular culture forms, as chapter 2 suggests, appear to work especially well for elders, but not always. Ned, for example, grew to dislike the "sameness" of *Murder, She Wrote* but had enjoyed traveling with the Starship Enterprise along the final frontier for almost three decades. For other elders, the parade of *CNN's Headline News* constitutes a daily ritual. What is common to all these experiences is the continual sense of cultural connection one feels.

In addition to its repetitiveness, Frankie had another complaint about the news—and about newspapers: Native Americans' viewpoints were never noted in public opinion polls. "We may be only 1 percent of the population, but it would be nice to acknowledge us from time to time." Feeling slighted did not discourage her from watching news, because she knew she had no other way of learning about events that interested her, such as government business, especially that involving the elderly. However, her sadness made her sorry for that dependence.

Like Rachel and Ayzik, Frankie was a troubled sleeper, but she did not watch television late at night for fear that she would wake her husband; for her, his comfort seemed a greater concern than her own. She preferred to read, both romance genre series and books by such popular authors as Jackie Collins and Harold Robbins. "He's nasty," she said. "He writes some really nasty stories. Some of the other authors do, too. But I know to expect it, and if it's a really good story, I'll read it all."

For about twenty years, Frankie had kept a neat list, organized by authors' names, in a single spiral notebook, showing all the books she had read. She had checked almost all of them out from the local branch of the public library, and the librarians knew to put aside books for her regularly. The notebook she kept presented her with a means to document her accomplishments in the wake of her frustration over not being able to advance in her job in a white-dominated state educational system. She preferred books to television, in part, because she felt unsympathetic to the Hollywood narrative about Indians: "We were always in the wrong. We were the ones scalping people, and we were the ones shooting them with our bows and arrows. But there had to be a reason we were doing this. But it always showed that we were the bad guys. Oh, we got some Indians rewriting history now."

Frankie said she made a point whenever she could to see a theatrical film that dealt with "new" ways of telling stories about Indians, such as *Dances with Wolves*. Preservation of her culture—especially in light of her growing grandchildren—was important to her, and she tried to share movies with them when she found it appropriate, just as she took her grandchildren to powwow each summer. Unlike Red Cloud and Wiley, Frankie found the classic Westerns too disturbing to tune out negative images in order to salvage nostalgia.

These Native American elders' tendency to read television against the grain—to discount the validity of the white image of the West, to filter out the prominence of the Lone Ranger with respect to Tonto—belies elders' reputation for "accepting" and "accommodating" positions toward television.[30] This reaction exemplifies Stuart Hall's theory that people who occupy marginal social statuses read media texts oppositionally; in this case, marginal status signifies a complicated interplay between social class and ethnicity and, in some cases, gender.[31]

Hmong Elders

Milwaukee's largest Asian population, gathered in poor neighborhoods mostly on its south side, is the thousands of Hmong refugees who settled there from Laos via camps in Thailand, where they had fled through the jungle during the Vietnam War. These displaced Laotians had suffered profound losses in behalf of the U.S.-backed cause during the war, and they risked invasion and chemical attacks after the U.S. departure from Indochina left them stranded in their homeland. Many died during the unsettled years, including close relatives of the four elders interviewed for this book. The refugees came to the United States—and to other sites, such as Australia—with little or no education and only spotty exposure to any culture outside their own; as an ethnic minority in Laos, the Hmong had been disconnected even from their national society. Yet religious conversion had taken place among many of the elders during their youth. Their families slowly abandoned the practice of animism in favor of Christianity, brought to them by Western missionaries. For many of the Hmong refugees, including the ones interviewed for this book, religious experience was a mixture of animism and Christianity, with the former seen as the more genuine identification and the latter seen more as a connection to Western success. For them, it was a matter of pride to be a Christian, because, for one thing, that meant one's children spoke English. The unity of the family was an essential marker of Hmong culture in America, where elderly men continued to play the role of patriarch and grandparents often lived in homes with unmarried children and the eldest son's family. For the patriarch, a life of leisure was expected, while elderly women generally cared for children and helped with other domestic labor while their daughters-in-law went out to work.

As with the American Indians, public programs serve free midday meals to Hmong elders at two senior centers on the south side, and that is where I located the three men and one woman who agreed to be interviewed with the help of translators. All of these elders had been sustenance farmers in Laos and received some form of public assistance in the United States, although this was not true for all the resettled Hmong elders. All lived with sons' families and in these contexts had seen television for the first time. Both the senior centers were located in extremely rundown sections, and the centers themselves exemplified this condition. Before lunch time, the Hmong seniors gathered in highly gender-segregated groups for conversation; none of them spoke more than a word or two of English. These were obviously very poor people. Some of them wore old versions of traditional Hmong clothing with brightly colored headdresses and shoes for the women. Many wore American-style garments that were poor fitting and had the appearance of thrift shop clothing.

All of the people who agreed to be interviewed in the two Hmong groups had come to the United States as much as fifteen years and as few as five years earlier. All were in their late 60s or early 70s. They now spent their days at the center and in their children's homes, where elderly women often were involved with child care and other domestic work and older men had little to do but did comparatively as they pleased. The woman we interviewed was Chin. Tung, Zhang, and Xiu were the men.

The Hmong elders reported different reactions to their first experiences with television, which had generally occurred when their younger family members purchased a set for the living room. "When I came to this country and saw the color television and everything, it was amazing," Tung said. "I am fortunate. I am very happy that every family in the United States can own their own television, any television that they can afford. I am pleased with the channels. They have so many different kinds that you can watch."

For Chin, the memory of not having a television had been a marker of her identity as a poor dirt farmer:

> You raise livestock, you just concentrate on that and the farm. You do not have the opportunity to attend school or go downtown to the lowlands so there is not the opportunity to see television. But you hear about it. When I come to this country, I see that there are many televisions everywhere you go. I feel upset and feel helpless that when I was young I had no opportunity to attend school so that I could know exactly what they are talking about on television here.

Chin not only had no English skills but had little knowledge, in fact, outside farm living and the daily life of a refugee. Years traveling through jungles and spent in Thai camps had not prepared her to decode news programs about social issues, trials, and taxes; mysteries; and game shows. Her children, who had acquired some education along the way and who spoke English fluently, were more successful at being American and thus at using television, she thought. She knew little or nothing about the kinds of programs her children watched. Whether family members had trouble finding work, whether a task was at hand in the kitchen, these were relevant matters to her. What the television movie was about generally was not.

Whereas the Russian elders I met had all relied on television to educate them about their adopted culture and instruct them in its dominant language, the Hmong elders had not. They saw television not as a resource so much as an insurmountable challenge. One reason for this seemed to be the Hmong elders' contentedness at remaining in the company of their own closed community; they lacked the Russians' passion for attainment of American citizenship, so there was no press to assimilate. They had, in fact, spent so many years as displaced individuals and came to the United States as a perhaps final step of displacement—not as a goal, like the members of the other two groups. Laos was far away, but socially, at least, the elder Hmongs attempted to remain there. Another reason for the Hmongs' difficulty with American television was that its basis was a language that was fundamentally strange to them. It may have been that the Russians found English easier to acquire because its structure was comparatively similar to Russian. (These particular Russians may also have had their language acquisition assisted by the existence of the Jewish Community Center as a more or less full-service cultural guide: upon coming to Milwaukee, they had been thrown together immediately with others in social settings where people who shared their ethnicity but not necessarily their nationality spoke English.)

The Hmong male elders said they understood only numerical figures when they saw these on television or perhaps "a few easy words," but Chin said she understood none of the content at all. All of the elders said they depended on someone else in the family translating for them, and, even then, they rarely picked up any depth of understanding. The exception was Zhang, who said he had become enthused about programs that show the possibilities for young people in the United States and programs that show how well the American system of government works. He saw these shows as a reflection of the optimism he had for his children, who had learned "how you pull yourself up" and who profited from education about government. "This political process is a gift," he said appreciatively. "In our country a leader is elected for life. In this country you have a choice for everyone to elect."

Zhang left open the possibility that the younger Hmongs would go back to their homeland: "The government election program is a good skill for them to learn and if in the future they have the chance to go back to their country, they will use that procedure."

In contrast, what the Hmong elders saw of depictions of aging on American television carried little resonance for them. "The family on television has a different living arrangement, a different attitude," said Tung. As Xiu said, "When they're in the nursing home, that's not applicable. For the Hmong, the old are old and the young have to take care of the old."

The theme that the Hmong elders picked up on despite their difficulty with language was a visual one. Like almost all the elders I have interviewed, they disliked what they found to be "disgusting" sexual and violent content in the programs they saw their children and grandchildren watching. These elders clearly believed that television was a powerful behavioral influence and could damage their traditional Hmong values and traditions. "Children learn from there very quickly," said Tung. "If they learn something from a film, pretty soon, it is imitated." Chin agreed, expressing fear that, although what she has seen has amounted to children imitating superheroes or teens imitating styles, eventually that might translate into young adults committing some kind of crime.

Tung pointed to the dangers of television's apparent glorification of adultery. "Some of the young married couples will think that it's OK to do that. They will start learning how to do that and find out life isn't like on the television."

Chin said she knew a young couple who had been married and divorced in America, and she thought the younger generation's continued exposure to immorality on television would only lead to more of such corruption in her Hmong society here.

Because of the language and other cultural barriers, programs that required no comprehension of English seemed important to the Hmong elders, who admitted that the sets in their homes were turned on a fair amount of the time (and were usually controlled by their sons when they were in the home). Tung and Chin said they both liked watching nature documentaries. "Some of the animals, I have never seen before," Chin said. "I enjoy watching these. You don't have to know the language, you can just imagine. You can see the picture movement."

The male elders said they especially liked watching nature shows that featured the scenery and animals they recalled in Southeast Asia. Monkeys were the favorite. Such programs presented rare opportunities to catch a glimpse of home, and seeing them was a bittersweet experience that resonated strong with the elders' foundation in animism.

The language barrier and cultural dislocation brought about a second favorite form of television watching—VCR tapes of Laotian and other Southeast Asian movies. Hmong families spend about twenty dollars each to acquire these tapes, which had been produced informally by contacts who had returned to Laos and bought master copies of them. Although many collections of tapes were small, the families borrowed them from one another. They would spend relatively large portions of their available cash to own them.

"I have the opportunity to see the background of the country," said Tung. "I enjoy watching those movies a lot, no matter how many times. Sometimes they are happy but sometimes they are sad and I cry."

For Chin, the movies were a reminder that Laos is a tiny country compared with the wide open space of her new world: "They show the homeland, the people. Some of the people that are on the movies, I recognize and I know personally. They may be my relatives. Some of the things in the movies are very close to my life, so they are effective. No matter how many times I watch, I still enjoy them."

Zhang said he liked seeing the ritual reminders that the movies brought back for him: "They show the religion, the ceremonies of the New Year, the popular singers."

Television constituted an embodiment of Western cultural expression for these Hmong elders. It remained fundamentally inaccessible to them and on some fronts repelled them but not when it was speaking their language. They constructed useful tasks for television in a make-do world. Television was doubly a dear memento and a grim reminder of a lost home. What seemed natural to these Hmong elders among television's content they embraced. What seemed unnatural they rejected. And while they did with television as they wished for themselves, they were fully cognizant of its potential bearing on their families' future. The medium remained a cavernous unknown, to be feared but not without reason. While these Hmong elders found television potently dangerous, they had been able to name the dangers they assigned to it despite their relative lack of knowledge about it.

For the Hmong elders, television obviously had narrowly defined roles, as they appropriated images for their socially instructional value and for purposes of nostalgia. We as media scholars like to think of television as a visually dominated medium, and it may be. But when people lack the necessary language skills to provide a context for most of the pictures, the medium is largely irrelevant to their lives. When they further lack the commitment to calling that culture their own, the prospects of their finding important places for television in their day are even more unlikely, as we can see by the Hmong elders' experience as distinct from the Russian elders'. Both the Russian elders and the Hmong elders found their ritual connections, in very different ways, as is evidenced, for example, by Rachel's and Ayzik's quest to learn English and by all the Hmong elders' repeated viewings of Laotian videotapes. Both groups exercised their particular passions for television in ways that could assist them in defining themselves as they traversed the blurred boundaries of double cultures.

Conclusion

The Practical Nature of Viewing

Because television rarely represents elders with any variety, we are left to assume the elderly as white and middle class (and generally male). This assemblage of elders conveys a substantial assortment of relationships between themselves and the medium. These relationships split messily along lines of group membership and whether or not one is a native English-speaker—two trajectories along which we can observe these particular "nomadic" subjects traveling.[32] I want to be clear regarding my intentions in using the term "nomadic" here and earlier in the chapter. I believe, as Radway and others have suggested, that we all occupy nomadic positions in the constant directional tacking of postmodern life. The Hmong refugees discussed in this chapter, because they have been physically displaced in recent years, feel this nomadic experience in a particular way, while the Native Americans feel it in another, related way. The Russian Jews have had yet another nomadic experience. But the gay and lesbian elders have nomadic experiences as well—and one that many other minority groups, including the three ethnic groups discussed here, share in their own particular ways. We are nomads as we move *through* culture and as the stuff of culture moves through us as ever changing social beings. Anthropologist James Clifford, who rejects the term "nomadism" because of its capacity for describing "natives" through a lens of colonialism, wants us to see culture as "sites traveled," as perhaps multiply located geographically. Cultures travel, Clifford notes, not just when diasporic movements of people result from contacts between their own cultures and others (as we see especially in the case of the Hmong people) but also when the products and media of a culture traverse it (across space and time) and give the people within it an experience of movement on a different level.[33] For example, Red Cloud, the Native American woman who liked watching television to monitor news events and to enjoy her revisionist telling of the cowboys and Indians tale, experienced television not only as a person who had been displaced (through space and time) but as someone who relied on television to travel back, away from daily life, to *places* she had identified as sites for understanding herself and her location in the world. Nomadic elder, traveling elder, Red Cloud pragmatically experienced mass culture as movement.

Despite my untidy efforts to report the overlapping strains on these elders' diverse positions, it is easy to see a common thread running through their stories: they each approach television on commonsensical terms. Like the public-access producers of chapter 4, they are conscious that many, if not most, of its messages are not intended especially for them, that, in fact, they are practically eavesdroppers listening in from some back room. Ultimately, they operate self-consciously from their perhaps marginal cultural position—whether they are nonwhite or non-English-speaking or gay, for example—when it comes to choosing television texts and establishing the relevancy of these to their own lives. And, I was surprised to learn, while television held little personal interest for some of them, none of them expressed vociferous anger. They were sometimes disappointed by its tendency to stereotype or overlook a category with which they identified, but they did not place wholesale blame on television for their marginal status. At the same time, they hoped for better things from a medium that they found to be powerful and potentially liberating for them in its image construction.

Television as a Social Learning Tool

Whether it was the gay or Native American elders expressing discontents or hopes about television's portrayals of particular minority groups, the Russians' self-styled cultural education, or the Hmongs' extrapolations from their children's mediated immersion into the American way of life, the elders I met shared a healthy respect for television's capacity to teach. In some cases, they greeted this perceived capacity with nearly unbridled fervor. Other times, they sensed doom in it. I sensed a third-person effect at work: like so many elders I have met (and nonelders, for that matter), these people thought that television's influence might be more powerful over others than over them; it was very much like the expressions made by residents of the Woodglen retirement community about the ills of commercial television and those made by *Murder, She Wrote* fans about the dangers of young people being exposed to televised immorality. The concern about fears and dreams for the young rang especially true among these minority elders. None of them theorized any sort of technological determinism, however. They saw with some clarity that there was wheat to enrich viewers and chaff to go to waste. Their varied responses to television contents also reminded me of work done by social scientists Marie-Louise Mares and Joanne Cantor on older viewers' responses to television portrayals of old age. Mares and Cantor found that elderly viewers' responses depended in part on their tendency to compare portrayals with their self-perceptions. For example, people said to be experiencing negative emotional states, such as loneliness or poor health, generally liked watching portrayals of elders that made them feel comparatively better off.[34] The elders in Mares and Cantor's sample did not so much look to television to teach in a blatant sense, as did the people I have written about in this chapter, but to help them make social comparisons and manage their moods. Both sets of elders, then, tended to see themselves as specialized consumers of television.

Championing Resonant Images

In connection with perceptions of television's power, the elders did not hesitate to cheer images that supported their perspectives. Most also vilified content that offended them. Some of the denouncements echoed statements of white and African American fans of *Murder, She Wrote*: television's showcasing of gratuitous sex and violence. An example from prime-time television illustrates how diverse the elders' opinions were, however. A long-running ABC situation comedy, *Family Matters*, was mentioned indirectly by two of the elders. Both were interested in the adolescent African American character Urkel, for quite different reasons. Urkel is a nerdy, bespectacled boy who is focused on his family to the exclusion of such matters as teen cool. Ned matter-of-factly volunteered, optimistically, that Urkel appeared to be a blossoming and sympathetically portrayed homosexual. By contrast, Rachel, who on occasion expressed anxieties about television's tendency to promote nondominant sexualities, expressed a liking for Urkel. She found him to be an altogether "typical American kid," a good boy. She would have been horrified by Ned's reading of Urkel's apparent sexual identity.

These three patterns—regarding practicality in viewing, seeing television as teacher, and using the medium to acclaim personally held views—are not exclusive to this array of minority elders. Yet they are patterns that seem especially serviceable to these elders in oppressed social positions. They used these approaches to negotiate meanings from television that made their own positions somehow more bearable. For example, Zhang found "the good" in television's capacity to teach his fellow Hmongs how to become favorably Americanized while retaining their ethnic identity. Byron and Cynthia prized television's progressive treatments of gays and elders because they felt confident that these would inform ordinary Americans about the groups to which they belonged. Television was in many ways trivial to these elders, but when it touched on matters that were untrivial to them, they took notice on behalf of their entire groups. It is unlikely that television will return the favor and take them so seriously.

These elders employed such approaches as practical viewing, using television as teacher, and affirmation of one's worldview as strategic means of staking out their personal positions in an ever changing media environment. They recognized, as did the public-access producers of chapter 4, that they were somehow an other to media culture, but this circumstance did not always seem evident to them with uniform clarity. For example, Ned ably participated as a fan of *Star Trek* because he chose to understand, on his own terms, at least, its long narrative as a postcolonial tale of a quest for understanding and for high-tech contact, not for imperialist domination. As a Trekkie, Ned found the program to be in line with his personal ideal of space travel, and he could enjoy the fantasy of somehow taking part. However, at other times, Ned felt the onus of other to roil his viewing experience, especially when he took television to be an instructional medium: when gays, blacks, or elders were portrayed in uncomplimentary or what he took to be unrealistic ways. What can we surmise when people such as Ned perceive themselves as standing along an interstice, fiercely clutching at some of television's political *visions* and spurning others? We are left to ponder Pierre Bourdieu's paradox: Can I resist what dominates me if I claim it for my own?[35] As the elders in this chapter move fluidly through the unfolding routes indicated by their strategic choices about television, I cannot supply a structuralist answer for this structuralist question. The best I can do is "sometimes yes, sometimes no." Obviously, it is one matter for people to locate places for ideological resistance and quite another for them to transform themselves materially. The sometimes acute political astuteness demonstrated by these elders and their measured regard for the effects of media have not eased their disenfranchisement. However, to the credit of many of them, they patiently maintain faith in television's ability to "play along" with change for the better, incremental as it might be, and they hope that television will do a better job departing from its practices of complicity with the dominant ideology that subjugates them. Television, for these elders, is a friend—but a friend to be watched.

Six

Theorizing Television in Old Age

This book has been an attempt to make sense of people's stories about the role of television in their lives. In pursuing this attempt, I have run headlong into the criticisms that such diverse audience-oriented scholars as Ien Ang and Lynne Joyrich have leveled at television studies in the ethnographic tradition. By making facile use of such social descriptors of age, class, ethnicity, gender, and religion, I have flirted with what Ang identifies as the "creeping essentialism" of audience scholarship, whereby one pledges to avoid generalization and then cannot manage to escape it because of the duty to make meaningful patterns from reams of interview transcripts and field notes.[1] For the audience scholar, the project of interpretation presents an irresolvable tension between such essentialistic writing and its counterpoint, unresolved specificity. As a poor solution, granted, I have tried to document, whenever possible, the slippery relations between the cultural categories that tend to be meaningful to us, not only as scholars writing about culture but as people moving through culture itself. What has resulted, and what I think is borne out among audience scholars publishing work in the ethnographic tradition, is not a compendium of formulas ("X age + Y class . . .") but a demonstration of how overdetermined subjectivities forge ordinary lived experience.

This project has meant getting at the stories people have to tell about subjects that I was curious about, and then interrogating those stories. When the goal seemed more attenuated, I relied on a narrower set of methods; when it was broad, the methodological choice was broader.

Chapter 2's probe of the role of a single television series in the lives of older women relied almost exclusively on textual analysis and group interviews to learn of people's positions. In the case of textual analysis, it was important to me to reflect on the televisual text of the murder mystery in order to situate it with respect to older viewers. In the homogeneous focus group, I reasoned, people could dynamically construct the terms that were meaningful to themselves—they could build knowledge together while the interviewer helped keep the group on path and shift the topic from time to time. (In order to find more individual texture, I talked separately with a few women later on.) I'm satisfied that, because this project represented a narrow research agenda, encouraging a group dynamic to help set the direction of the "answers" was a useful idea. But it was not the only idea I could have invoked, and it might have been more satisfying for the reader to have been able to "experience" the fruits of field observation, to learn more about these women's home circumstances. I regret that my methodological choice preempted such experience.

Likewise, chapter 5 relied almost totally on depth interviews with some inclusion of observations about domestic life. What might have happened if such a study, about the broad role of television in the lives of minority elders, had involved visiting these elders in their homes over a period of months instead of limited contact through interviews, a few of which were not, for practical reasons, conducted in the homes of the elders? No doubt, a richer tale would have emerged. In chapter 4, interviews and observations, supplemented by textual analysis, seemed an obvious choice to learn about public-access production by elders, but again, other choices might have yielded other rich material. I am happiest with the methodological choices I made for chapter 3, the study of the retirement community. During the two years I worked on my dissertation, I luxuriantly engaged in repeated observations, group interviews, and depth interviews that made me feel I understood these elders generally better than the ones I got to know subsequently in Milwaukee. In this case, the job was constructing the narrative of a community, its discursive structures, and television's fit in its sometimes hard-to-identify spaces. A wide range of methods seemed sensible.

Relating stories about the currency of television in the lives of a diverse class of people—such as the elderly—meant, for me, the strategic sacrifice of the "perfect" methodological formula in my attempt toward inclusiveness. If I felt I could, I would still be conversing with my volunteer "subjects" and visiting their homes. I would have jealously guarded those moments from graduate students who generously assisted me and were able to relish their own moments with these elders. The practicality of pressures external to the work—including the scholarly need to publicly understand contemporary issues with some expedience in order to make that understanding useful to others—necessitated choice making.

We talk sometimes of the "practical" project of cultural studies. In doing so, we tend to focus on the end rather than the means. To be sure, we should have some goal in mind when we go to the trouble of bothering people with pesky questions about the intimate details of their lives. It is not enough, morally, to claim the broad banner of scholarship. To be sure, I will make every attempt to circulate the ideas in this book to benefit further a public goal of understanding the aged's lives with television. It is insufficient that a community of scholars might respect elders' relationship to media; it seems imperative that television's creative workers, gerontological workers, and older people themselves have a chance to ruminate about it.

There is, obviously, another practical side of cultural studies, however, and that is the means of engaging in it, as demonstrated by my confessional tale about methodological choice. What I have done here is no pretense for ethnography but a sheepish apology for the lack of it. Snippets about television audiencehood channeled into my ideas about social class and such factors as age-related health. Such constructions, as Ang and Joyrich admonish us, tend to reify social categories. Because we constantly work at knowing the place of the media scholar in the academy, such methods of description also tend to reify the primacy of television and its place—among other media and amid everyday circumstances.[2]

I have been aware of making strategic, practical choices in order to get at the persistent question "How do we talk about television in the lives of the elderly in American society?" My practical goals have steadied my hands as I have held the tape recorder and tapped at the keyboard even as I am haunted by the unanswered plea made by Janice Radway in which she critiqued audience studies so effectively. Radway cogently argued against the artificiality of scholars' attempts to understand the role of media by engaging in methods as narrow as the ones used for this book and for her own landmark study on romance readers. She called for structural changes to the academy itself that would encourage team approaches to interrogating the place of leisure itself—generally and specifically—within a whole community. Only when we know the fit of leisure in a place—and deeply understand that place and its people—can we fathom the place of television watching in the lives of a particular group, or some similar issue.[3]

A decade later, Radway's call has gone largely unanswered, the academy remaining systemically user-unfriendly. (The system of faculty rewards and punishments, if anything, has effectively become more alienated toward the conception of any such comprehensive "team" ethnography.) What we have instead are modest and sometimes marvelous beginnings. Little sketches such as those in this book give us textured maps of meaning that might never add up to general understanding—the robustness of social science plus the depth of cultural studies—but that give some sense of how people's voices sound.

While I won't generalize (a phrase that we audience scholars fetishize and inevitably abuse, so much so that it has come to seem like the cliché of a comedy routine), I will say some things about this group of seventy elders as a complement, who steadfastly countered the negative stereotypes we have come to recognize as mythic Western truisms:

- First, these elderly people are both aware and attentive with respect to their relationship to television and what it offers them. Despite their wide range of ages and other bits of background, they demonstrated, on the whole, a thoughtfulness, a purposiveness, an awareness, a clarity, a poise, and, in most cases, a zest with respect to life in general and television specifically. Without fail, they displayed a readiness for my questions, and many were ready with questions for me. I never heard one of these elders utter in response to a question, "Well, I don't know" or "I haven't considered that" or "I'll have to think about that." They all showed a surefootedness in describing their reflections.
- They had been profitably seasoned by the seasons. On the whole, a mellowness seemed to prevent them from angering at television even if they felt critical of it. I got a sense of people realizing that, even if they wanted to blame television for their own problems or for society's, it made little sense to become enraged. They seemed to have acquired

some perspective, some personal distance, that we younger folks sometimes lack. Of course, there were exceptions to this. Ayzik, the Russian elder, raged about television's inclusion of homosexuals, which he found outrageous and which seemed to upset him tremendously. In the group interview setting itself, spirited arguments broke out between women a couple of times, and I found myself embarrassed by the confrontational styles of those involved. (Not that I don't have my own confrontational style, but here I was, the interviewer.)

- These elders seemed to mine television for what seemed the genuine artifact to them. This is not to say that they always favored what might be labeled the "best" television shows, but they did seem to share a thirst for programs that would appeal to their sense of what mattered, what seemed real. Sometimes, this meant the display of values that they found enduring. Other times, this meant a prejudice against fictional content or what they might have found to be confounding production styles. For some, this made the stark elegance of British mystery programs appealing. For others, it made game shows or political conventions attractive. These elders tended to tell me that the so-called conspiracy dramas, such as *The X-Files*, smacked of poorly conceived revisionist storytelling; such programs were decidedly unpopular among the elders, as had been the more entropic dramas such as *Twin Peaks* and *Moonlighting*. Such work with respect to television viewing bolsters the statement made by Wilbur Schramm three decades ago: mediated communication fosters social participation among the elderly and deters their disengagement from society.[4]

The relative poise, mellowness, and discriminating eye that these elders demonstrated made me self-conscious about my own mercurial tendencies. In my haste to slot them, they forced me to slow down and reconsider the possibilities. As much as I fought the tendency to deny it, I had somehow expected the richer ones to be smarter, the "different" ones somehow to be exotic others. Instead, these elders reminded me of what I already knew on a deeper level—that wisdom is unevenly acquired, even within an individual's own knowledge base. Among the most poorly educated of my spotty "sample" were Ned, Cynthia, and Vercella. Among the best educated were Barbara, Lila, and Mildred. I learned plenty from each of them, and I figured that there was plenty that each, like me, would never learn.

When I began studying older people, I felt comfortable with the idea because I thought I had always implicitly understood them. I have seen my own elders move along continua of autonomy and dependence, wisdom and naïveté, acceptance and denial. But although I longed to know the diversity of elders, I was unprepared to articulate it. Struggling to do so has made me only more curious and hopeful of understanding.

Whether it is younger relatives with "active" lives or social scientists with a need to categorize populations neatly, Americans have linked television viewing among the aged with deadly passivity, hinting that elders who spend hours each day with the tube have ceded over their lives.[5] This argument has raged, from a scholarly perspective, for three decades and showed little development in a recent study published in the *Journal of Applied Gerontology*: "The greatest portion of the impaired elder's day seems to be characterized by passivity, notably as reflected in the euphemistically named Rest category and very probably in the television/radio category. Although some portion of television watching is unquestionably active and stimulating, some is just as certain to be unattended or simply background stimulation." The scholars go on to generalize, abandoning the notion of "stimulating" television: "Together, rest and television accounted for over half of the day (55%), followed by another 17% of additional passivity when the elder was the recipient of help by others, and another 9% spent in pathological, perseverative behavior."[6]

The implications of such remarks are clear: elders ought to be *doing* something with their time. In the words of the often aired Centrum vitamin commercial for older customers: "It's a great time to be silver!" On these terms, television is at best wasteful, at most dangerous. It's a treacherous sea that one traverses while watching television in old age. There is shuffleboard to be played, investments to be made, after all. In keeping with most of television advertising and entertainment's portrayal of people with "active" lifestyles, the vitamin hawkers at Centrum show their seniors smiling while playing sports and dancing, concentrating on the realm of physical activity. In doing so, they ignore the very active cognitive "play" of watching television that their ideal vitamin takers might choose as well. Anxieties about television's mode of forced passivity have become so ingrained that it seems laughable for an advertiser to argue that audiencehood might mean active engagement. (This is ironic considering the advertiser's fervent hope to make a sale by competing against like companies to use television rhetorically, in order to convince viewers of a product's worth; does that make viewers passive or active?) Centrum's implication to the home audience of elders (and extending to the home audience of nonelders) is that if you're going to be old, you'd better "get Silver." Otherwise, you're merely Gray. Old people who want to amount to something had better play golf or do the fox-trot.

Betty Friedan sharply criticizes the installation of the Sun Belt retirement village lifestyle because, in part, it throws people together on the basis of cohort and causes them to relate to one another as people who have nothing left but old age and the passion to run from it. This communal placement of elders, Friedan argues, underscores their already marginal status.[7] Friedan despises this insistence on distraction from aging, preferring instead to see old age as a locus of personal wisdom and strength. This sometimes artificial "active senior"—read "middle-aged" retiree—who is the ideal product of so many canasta games and bus tours may indeed be a wonderfully fulfilled human being. But this is not the life for all elders, or perhaps even most. In fact, journalists in recent years have been producing accounts of a growing phenomenon that they apparently find fascinating, almost in a man-bites-dog sort of vein: retirees are intrepidly leaving the suburbs and settling in big-city apartments, especially in New York, to take advantage of easy transportation, cultural amenities, and other benefits. These retirees are often quoted as saying that they find cities "more real" than retirement villages, that they are looking for a more meaningful experience. It is difficult from outside old age to judge what is meaningful and what is not, and television use is no exception. The elders I have met in connection with this book have taught me, on the whole, that something significant is occurring when an elder may be continually engaged with television, day in and day out; what has been popularly regarded as a catatonic absorption of meaningless fog cannot be generalized to the larger elderly audience experience, which can have no unilateral explanation.[8] And, as we have seen in so many ethnographic studies of younger people, the "act" of viewing never occurs but in a complex interplay with other events and texts.

Social scientists have explained what some of these purposes are that elders traditionally have liked to apply to television: companionship and security, cheap entertainment, intellectual stimulation and challenge, social surveillance, and the like.[9] What we need to do as scholars—and, more generally, as outsiders to old age—is to begin to validate some of these needs that elders obviously need to fulfill. Then perhaps we can respect their choice for selecting what many of them consider the best available provider for these needs. Such validation, of course, is unlikely to result except following the more basic cultural validation of the elders themselves.

Some clues to unriddle the means to validate our elders have already been scattered for us. For example, a 1994 survey found that older Americans are substantially more likely than younger people both to tune in to television for a specific show—purposive viewing—and to express dissatisfaction with the choice of things they can see on television.[10] Many other such clues, I trust, have been scattered throughout this book. Such evidence does not support enduring stereotypes about a docile aged public. It implies instead creativity, particularity, and an insistence on taking oneself perhaps more seriously than others take one.

Sometimes creativity is hard to validate from an academic position. Here, we harbor middle-class biases that we unleash with more brutality than do the Centrum-style popular critiques against elders and other nondominant classes. This book, for example, has interpreted with breathless excitement the more or less political engagements of elders with mundane murder mysteries and other entertainment programming while prizing their hard-edged focus on middle-brow information outlets such as C-SPAN and community-access production. In the process, I have glossed over such practices as ritual engagement with game shows and certain sports events, to which many elders flock not out of concern with participation but because of the shows' docile curb appeal.

I know of one woman in her late 80s, with a grade school education, who counts among her favorite programs *Wheel of Fortune*, *Jeopardy*, and certain sports broadcasts. To my knowledge, this woman, Reneè, has never tried to solve the *Wheel of Fortune* puzzle. She recognizes a show as a rerun because she recalls a contestant, not a puzzle. I do not believe she has ever shouted out a *Jeopardy* answer (a question?). She is a fan of faces in these shows. Vanna is glamorous, Pat sincere, Alex self-assured. She roots for the "nicest" contestant—the housewife with four children or the accountant who wants to use his winnings to take his wife on an anniversary cruise. She balks at contestants who "get greedy"—wagering and losing money when they had safely laid claim to a respectable amount. But the *Jeopardy* definition signified in "Famous Sonatas for $1,000" both eludes and is irrelevant to her. Her experience with watching baseball or golf is similar. She may express an awareness that home-run hitter David Justice is in a "slump," that golfer Tiger Woods has "birdied all afternoon"—she has heard an announcer make these statements. Yet Reneè has no clue what a baseball "hit and run" might be or that a golf tournament has rounds that do not get telecast. She likes baseball because it is slow and fairly quiet and has a pastoral setting: the field is lovely and neat, the pro-family crowd shots endearing. Men's golf, framed by green vistas and hushed commentators, lures her with frequent shots of tidy young men and earnest-looking female relatives. This woman loves to talk about her television viewing with family members and friends. In a world that she has come to recognize as increasingly unsettling and unhappy, her manipulation of TV content choices brings her tremendous satisfaction. Her experience demonstrates a creativity that is as valid for her as another's plugging into C-SPAN to make political choices. I'm not equating creativity with citizenship—I think both are important. But Reneè is mostly uninterested in citizenship issues. If her choices might be painful for me to watch in some ways, I respect her resourcefulness.

This book has, on the whole, celebrated the exercise of citizenship by elders, although it has been a sort of vulgar citizenship that I have sometimes elevated. I have had to remind myself many times that I cannot mistake serious-minded viewing choices for cultural power. Most of the people whose stories I have told consider their own public participation within existing outlets prescribed by existing media frames in liberal capitalism. They used these frames to unload for me their criticisms of television's impact on them as citizens; in doing so, many of them were able to appropriate television's texts to fit their own sympathies. For example, Frankie, who railed against television news programs' shirking of coverage of Native American opinion, did not take her complaint "public," writing to networks or joining some boycott effort. Like most of us, she complained as a sort of aside to her practice of continuing to watch the very programs that disappointed her. She did make frequent efforts to take her grandchildren to see movies that she felt most accurately portrayed Indians. And she urged her children to accompany her to annual powwows and worked with an agency to get a home built for poor Native American elders. In the spaces where they found the exercise of citizenship comfortable, elders such as Frankie toiled. Knowing that few younger people took them seriously bothered them but did not paralyze them. These elders expressed their personal- and civic-focused energies in remarkably diverse ways. But even when they wanted television to help them seek relief in a life that had made them feel tired or weak, they worked to reconcile the texts they sought out with their own perceived positions in a morally correct life. These elders belonged to cohorts that had known the Great Depression, World War II, and the New Deal, in some way or other. They grew up in times when a conservative morality, teamwork, self-sacrifice, delayed gratification, and compliance with authority and conformity were honored practices. They saw lean times turn into economic expansion and have seen recesses in that expansion. Many of the most senior of these elders remained quite frugal while the younger ones grew accustomed to spending but resisted the practice of material disposability.[11] Ultimately, their approach to television, no matter how trivial it might have seemed from the outside, was meant to be sensible and morally correct, two supreme values among these elders.

Baby boomers, once they begin to reach the category of old age—after no doubt resisting and displacing its operating definition—surely will teach us different lessons about how media should be used. For one thing, the pressure from (and on) baby boomers both to prize youth and to succeed at old age will continue to elongate middle life and to forestall the age of physical decline and mortality, by whatever means possible. I believe this process will help underscore television's increasing ambivalence about the aged—affection for the old who appear young, denials about the old who are lonely, sick, and decrepit.[12] What values old boomers will wish to validate will focus more on the areas of individualism and personal control of their social environment. Television, in whatever incarnation it is to be experienced, will offer a far less stable field of meanings in these elder lives. At this point in the coming years, postmodernism's harshest (or most liberating) impacts will perhaps be finally linked with the slippery, unstable "elderly audience." I have tried to show that television in its current state, together with the so-called elderly in theirs, represents in large measure a multidetermined field of meanings, but it is apparent that aspects of this field appear quite stable, such as a preference for a certain moral component in television. Today's elderly appear, then, to experience television both as a stable force and as evidence of the postmodern condition, and these positions are, as we might imagine, often difficult to accept simultaneously. I have spoken with many older women who were drawn to *Murder, She Wrote* for its purity, for example, but they were also drawn to the lurid content of daytime talk shows. They wanted to spurn and sense contemporary life at once, to occupy high moral positions and experiment with nefarious ideas. Ultimately, they felt the weight of their conservatism tug at them more as they expressed greater comfort with meanings that appeared time-tested, sturdy, and chaste. As these elder generations—television's first viewers and first audiences of retirement age—begin to disappear, the field of television-elder relations undoubtedly will change on this important front and on many others. It will be an enormous task to begin understanding this shift.

Notes

Preface

1. Betty Friedan, *The Fountain of Age* (New York: Simon & Schuster, 1993), 49.
2. Margaret Clark, "The Anthropology of Aging, a New Area for Studies of Culture and Personality," *Gerontologist* 7 (1967): 55–64, and quoted in Barbara Myerhoff, *Remembered Lives: The Work of Ritual, Storytelling, and Growing Older* (Ann Arbor: University of Michigan Press, 1992).
3. Myerhoff, *Remembered Lives*, 101.

One Television and the Elderly Audience

1. Of course, the commercial works on a different level, too, in its insistence that stereotypes about elders are beginning to fall away. We see through the image of the independent, active older man in the commercial the truth that Americans of retirement age may be starting to reject the mantle of "elderly."
2. See, for example, Nancy Wood Bliese, "Media in the Rocking Chair: Media Uses and Functions among the Elderly," in *Inter/Media: Inter-personal Communication in a Media World*, ed. Gary Gumpert and Robert Cathcart (New York: Oxford University Press, 1986), 573–582.
3. John Tulloch, "Approaching the Audience: The Elderly," in *Remote Control: Television, Audiences, and Cultural Power*, ed. Ellen Seiter, Hans Borchers, Gabrielle Kreutzner, and Eva-Marie Warth (London: Routledge, 1989), 180–203.
4. Ien Ang, *Desperately Seeking the Audience* (London: Routledge, 1991); John Hartley, *Understanding the News* (London: Methuen, 1982); David Morley, *Television, Audiences, and Cultural Studies* (London: Routledge, 1992); and Janice Radway, "Reception Study: Ethnography and the Problems of Dispersed Audiences and Nomadic Subjects," *Cultural Studies* 2, no. 3 (1988): 359–376.

5. See, for example, Richard H. Davis and James A. Davis, *TV's Image of the Elderly: A Practical Guide for Change* (Lexington, Mass.: Lexington Books, 1985).
6. Jenny Hockey and Allison James, *Growing Up and Growing Old: Ageing and Dependency in the Life Course* (London: Sage, 1993).
7. Kathleen Woodward, *Aging and Its Discontents* (Bloomington: Indiana University Press, 1991), 193. Barbara Myerhoff, in *Remembered Lives: The Work of Ritual, Storytelling, and Growing Older* (Ann Arbor: University of Michigan Press, 1992), asserted that "literate, complex, Western, industrial societies penalize old age most heavily," because in such societies, the meaning of old age is surplus (122).
8. U.S. Bureau of the Census, 1990.
9. Jeff Ostroff, *Successful Marketing to the 50+ Consumer* (Englewood Cliffs, N.J.: Prentice-Hall, 1989).
10. M. L. Joyce, "The Graying of America: Implications and Opportunities for Health Marketers," *American Behavioral Scientist* 38, no. 2 (1994): 341–350.
11. One place where this scene appears to be shifting is the Internet. As older computer users increase in numbers and more go on-line, such Web sites as AARP's are getting more traffic and, presumably, building attention. AARP's Web site, along with others geared toward elders, offers frequent focuses on organized political efforts, such as congressional letter-writing campaigns built around age-related issues such as health care or Social Security reform. The growing prominence of publicity sprung from an Internet platform may either cause the conventional news media to turn more attention to such groups as AARP or make their oversight less relevant.
12. For examples, see Janice Radway, *Reading the Romance: Women, Patriarchy, and Popular Literature* (Chapel Hill: University of North Carolina Press, 1984), and Ann Gray, *Video Playtime: The Gendering of a Leisure Technology* (London: Routledge, 1992).
13. Robert O. Hansson and Bruce N. Carpenter, *Relationships in Old Age: Coping with the Challenge of Transition* (New York: Guilford Press, 1994).
14. All the participants were over 60 except for one woman, who was 56. I had initially intended to interview both her and her mother, but after several encounters in which it seemed that this would be viable, the mother's unavailability became evident. The 56-year-old's account of her own experience was so closely linked to her mother's and so compatible with the accounts I received from others in similar situations that I decided her story was valuable enough to include.
15. Seiter et al., *Remote Control*, 2.
16. Elihu Katz, "Introduction: The State of the Art," in *Public Opinion and the Communication of Consent*, ed. Theodore L. Glasser and Charles T. Salmon (New York: Guilford Press, 1995), xxi-xxxiv.

Two The Case of the Mysterious Ritual

1. Older women and men as well as middle-aged women frequently watch suspense and mystery programs, according to George Comstock et al., eds., *Television and Human Behavior* (New York: Columbia University Press, 1978). My ethnographic

research in a midwestern retirement community uncovered a keen interest in classical mystery and detective dramas among elderly women there.

2. See Raymond Williams, *Marxism and Literature* (Oxford: Oxford University Press, 1977).
3. Ritual is "an act or actions intentionally conducted by a group of people employing one or more symbols in a repetitive, formal, precise, highly stylized action." See Barbara Myerhoff, *Remembered Lives: The Work of Ritual, Storytelling, and Growing Older* (Ann Arbor: University of Michigan Press, 1992), 129. Ritual can be an especially meaningful component of everyday life for an older person, who may feel caught in the absence of and with the need for acts that help mark the passage of life experiences in old age, Myerhoff argues.
4. Myerhoff reasoned that rituals, because of their "repetitiveness, formality, and rigid precision," can construct their "fictions," or stories, around "dangerous" themes (ibid., 130).
5. Roger Silverstone, building on Victor Turner's theorization of ritual and the liminal place of it in people's lives, suggests that ritualized television viewing leaves us in this in-between space. See Silverstone, *Television and Everyday Life* (London: Routledge, 1994).
6. Horace Newcomb and Paul Hirsch have contended that television functions as a cultural forum and that people use texts to ritually connect themselves to their cultural traditions and changes. Horace Newcomb and Paul M. Hirsch, "Television as a Cultural Forum," in *Television: The Critical View*, 5th ed., ed. Horace Newcomb (New York: Oxford University Press, 1994), 503–515. Newcomb and Hirsch have pointed out that much of television encourages ritualized viewing, with people being drawn to the same programs in the same time slots week after week, day after day.
7. John Cawelti, *Adventure, Mystery, and Romance: Formula Stories as Art and Popular Culture* (Chicago: University of Chicago Press, 1976); Jane Feuer, "Melodrama, Serial Form, and Television Today," in Newcomb, *Television*, 551–562. Rituals, both sacred and secular, communicate important symbolic information, but they are not messages that are free of problems. Often, rituals allow people to shun direct confrontation with significant, perhaps frightening, events, acting as bridges as they transport participants across rough waters. The viewing of murder mysteries, in which characters represent the repetitive interplay of good versus evil and evil goes explored without punishing the viewer, allows such bridge crossing. See Richard Schechner, *The Future of Ritual: Writings on Culture and Performance* (London: Routledge, 1993).
8. Myerhoff noted in her work with elders a tendency to storytell that allowed people to focus on the core values of their lives and the stability of their relationships with one another. Barbara Myerhoff, "A Death in Due Time: Construction of Self and Culture in Ritual Drama," in *Rite, Drama, and Festival*, ed. John MacAloon (Philadelphia: Institute for the Study of Human Issues, 1984), 149–178.
9. Tim Brooks and Earle Marsh, *The Complete Directory to Prime-Time Network TV Shows: 1946–Present* (New York: Ballantine Books, 1982).
10. According to Alex McNeil, *Total Television: The Comprehensive Guide to Programming from 1948 to the Present*, 3d ed. (New York: Penguin Books, 1991), *The*

Rockford Files was number 12, *Mannix* was 19, and *Cannon* and *The NBC Sunday Mystery Movie* tied for 20. Other programs oriented toward the mystery formula in the top twenty included *Kojak* and *Hawaii Five-O*. *Kojak*, *Hawaii Five-O*, *Mannix*, and *Cannon* all were CBS series.

11. Created by Richard Levinson, William Link, and Peter S. Fischer, the series premiered with CBS's fall 1984 lineup, when star Angela Lansbury was 59 years old. It quickly became a network staple, dependably winning its time slot and generally finishing in the Nielsen top ten.
12. "Lansbury's 'Jessica' Is Why 'Murder' Is a Decade-Long Hit," Associated Press (November 11, 1994), appearing in *Milwaukee Sentinel*, 3C.
13. Paul Farhi, "On TV, Madison Avenue Sets Dial for Youth," *Washington Post*, September 16, 1994, A1.
14. A thirty-second commercial could be had in the fall 1994 season on *Murder, She Wrote* for $116,000. This series placed number 16 in the May 1994 sweeps rating period for prime-time programming, substantially higher than 94th-place *Lois and Clark*, whose viewers skew much younger. A thirty-second spot on *Lois and Clark* cost $132,000 in the fall of 1994. See Betsy Sharkey, "The Secret Rules of Ratings," *New York Times*, August 28, 1994, B1.
15. Robert Masello, "That's All She Wrote," *TV Guide*, November 4, 1995, 26–29.
16. The very next season, CBS executives admitted that dumping the elder market had cost them, and they lured such viewers again with such shows as *Touched by an Angel*.
17. USA Network bumped *Murder, She Wrote* from its evening schedule in the summer of 1997, but the Family Channel, another outlet whose audience skews toward older people, planned to pick it up.
18. Lyn Thomas, "In Love with Inspector Morse: Feminist Subculture and Quality Television," *Feminist Review* 51 (1995): 1–25.
19. Cawelti, *Adventure, Mystery, and Romance*.
20. All these programs bear a resemblance in form to the traditional drawing-room mystery found in literature, as typified by Agatha Christie in the 1920s and 1930s. For a more detailed explanation of the mystery formula, see ibid. and M. A. Collins and John Javna, *The Critics' Choice: The Best of Crime and Detective TV* (New York: Harmony, 1988).
21. In the made-for-television *Perry Mason* movies, set in the late 1980s and early 1990s, characters reflect changing times. Della Street acts less deferentially to Mason, often winning her boss over to her view. The two private detective characters, including the first, Paul Drake Jr., act with more recklessness and worldliness than the original, flat Paul Drake showed. The first Paul Drake wore light-colored suits, smoked cigarettes, and commented wryly on his investigations; the latter detectives got into fistfights, car chases, and messy relationships.
22. Collins and Javna, *The Critics' Choice*, 29.
23. Patricia Mellencamp, *High Anxiety: Catastrophe, Scandal, Age, and Comedy* (Bloomington: Indiana University Press, 1992), observes the appealing combination of Jessica's status as both an independent sleuth/world traveler and a small-town citizen.
24. See ibid. for comment on the program's preference for rationality over affect.

25. See, for example, Ellen Seiter, Hans Borchers, Gabrielle Kreutzner, and Eva-Marie Warth, "Don't Treat Us Like We're So Stupid and Naïve: Toward an Ethnography of Soap Opera Viewers," in *Remote Control: Television, Audiences, and Cultural Power*, ed. Ellen Seiter, Hans Borchers, Gabrielle Kreutzner, and Eva-Marie Warth (London: Routledge, 1989), 223–247.
26. For example, Fiske posits the quiz show as an enactment of capitalist ideology through its rituals of emphasizing the personal differences among the competitors in the introductory segment and establishing the triumph of the winner at the end through ritualistic celebration that includes emphasis on material prizes. See John Fiske, *Television Culture* (London: Methuen, 1987).
27. Klein, in making a case that detective stories tend to support male hegemony, has suggested that it is significant when a woman detective lacks the official status of detective we most often see connected with male detectives: "When detectives are amateurs, they can be ignored and their behavior seen as a momentary intrusion into public life. And, the changes in social organization which would arise from women's active participation in public life, disruption of economic activity, and involvement in the political process could be dismissed as short-lived and inconsequential." See K. G. Klein, *The Woman Detective: Gender and Genre.* (Urbana: University of Illinois Press, 1988).
28. Rob Edelman and Audrey E. Kupferberg, *Angela Lansbury: A Life on Stage and Screen* (New York: Birch Lane Press, 1996).
29. Ibid., 217.
30. Most subjects were members of either the Christian or Jewish faith. Much of the data came from homogeneous focus groups, but I conducted long, semistructured interviews with several people to get more detailed information. Most research subjects lived in metropolitan Milwaukee, including urban and suburban areas, but a focus group of five women comprised rural working- to middle-class women living in northern Wisconsin. I interpreted class affiliation based on education levels and occupational ties.
31. About twenty-four hours of interview and focus group sessions were taped for the study with the assistance of graduate students.
32. People and place names have been changed to ensure anonymity of participants.
33. I tried to let participants direct the tenor of the conversation about the show. I asked general questions such as "What do you like about the program?" and, as specific topics developed, asked them to elaborate. In most of the focus groups, it was important for the interviewer to speak often to prevent one or two participants from dominating the discussion.
34. I am using the term "interpellation" in the same way intended by Louis Althusser and media theoreticians who have adopted it to convey the mode of address that television uses to engage audience members. *Murder, She Wrote* uses devices such as the friendly, uncomplicated demystification of new technologies to interpellate audience members who welcome that sort of demystification.
35. See Thomas A. Lindlof, "Media Audiences as Interpretive Communities," *Communication Yearbook* 11 (1986): 81–107; Janice Radway, *Reading the Romance: Women, Patriarchy, and Popular Literature* (Chapel Hill: University of North

Carolina Press, 1984); and Henry Jenkins, " 'Strangers No More We Sing': Filking and the Social Construction of the Science Fiction Fan Community," in *The Adoring Audience: Fan Culture and Popular Culture*, ed. Lisa Lewis (London: Routledge, 1992), 208–236. Radway did not construct the romance novel audience as an interpretive community but clearly established its members in this way.

36. Biographical sketches of subjects were drawn from their answers to brief questionnaires administered prior to interviews.
37. When the trial ended, the reaction of elder African Americans was enthusiastically favorable. About 40 percent of attendees are white. These people silently waited for the bus home, and some did not come to the center the following day.
38. Among the African American women, Ben Matlock was also appealing because he was a southerner and because he had starred in *The Andy Griffith Show*, based in the small-town South.
39. In an informal interview I did with a men's club at the Jewish Senior Center, I learned that several of the members enjoyed watching *Murder, She Wrote* with their wives. Freedom from sex, violence, and rough language was the primary pleasure they said they took from the show. Among these particular men, Jessica Fletcher was a popular heroine because of her autonomy as an aged person, they told me.
40. For elaboration, see Ellen Seiter, *Sold Separately: Parents and Children in Consumer Culture* (New Brunswick, N.J.: Rutgers University Press, 1993).
41. Helen and Bea may have attributed this ability to forget "whodunit" to age, but I know many mystery viewers, myself among them, who experience the same thing.
42. Mellencamp, *High Anxiety*, 315.

Three Television Use in a Retirement Community

1. George Comstock et al., eds., *Television and Human Behavior* (New York: Columbia University Press, 1978).
2. In fact, many cases exist in which households are too limited a category for study. For example, in college dormitories, where residents spend much of their time in close contact with one another and sometimes share viewing areas with many people, it is difficult to distinguish the meaning of the "household." The interpretation of television is embedded in the dormitory suite and in the larger space of the residence hall. Residents leave the dorm periodically to rejoin their families at home, where interpretation also is embedded.
3. As David Morley and Roger Silverstone have suggested, media are integrated into these domestic and communal contexts, and these are the sites of interpretation. First, then, it is necessary to study Woodglen residents as domestic subjects, people incorporating media into their home in the everyday sense. For example, gender relations influence how media are used and by whom. In the case of aging couples, the retirement of one spouse may spark a change in gender relations with respect to control over the television set at various times of the day. In other cases, the death of a spouse may change how a surviving spouse perceives her or his relationship to television. Class position, educational level, region, and race also inform media experiences. Factors of aging, such as membership in generational

cohorts and age-related physiological changes, may produce changes in how television is used in homes.

4. Arlie Russell Hochschild, *Unexpected Community* (Englewood Cliffs, N.J.: Prentice-Hall, 1973).
5. Pierre Bourdieu, *Distinction: A Social Critique of the Judgement of Taste*, trans. R. Nice (Cambridge: Harvard University Press, 1984), 32.
6. Ibid., 33.
7. Charlotte Brunsdon, "Crossroads: Notes on Soap Opera," *Screen* 22, no. 4 (1981): 32–37.
8. Herbert Gans, *Popular Culture and High Culture: An Analysis and Evaluation of Taste* (New York: Basic Books, 1974).
9. For a social explanation of retirement communities, see H. R. Stub, *The Social Consequences of Long Life* (Springfield, Ill.: Charles C. Thomas, 1980). Even in retirement communities, distinctions are to be made about the meaning of leisure, because of disparate economic and cultural capital found among these. In one community, shuffleboard courts, canasta parties, and ceramics classes form the program. In others, yachting, tennis courts, and golf courses provide the means for recreation.
10. One of Barbara's neighbors, Fostine, was among the "fortified Bush people" who objected to the petitioning. She mentioned in a focus group interview that she had found it "unseemly" that a resident had carried a political petition into the dining room. Later, in an interview, she told me she thought Barbara's apparent support of Perot to be "an expression of disloyalty to the president." Another Republican neighbor, Verna, expressed agreement in a subsequent interview. It did not seem to matter to either woman that Barbara might vote for someone other than Bush, but to publicly introduce opposition to the two-party system, they thought, conveyed meanness toward the incumbent.
11. Benedict Anderson, *Imagined Communities* (London: Verso, 1983).
12. One of the hazards of doing self-reported research is the risk the researcher takes in having the data shaped by the subjects' perceptions of what is wanted by the researcher or by how the subjects wish to make themselves seem to the researcher. I do not know that people at Woodglen did not watch situation comedies or television movies for escapist reasons, but I can say that an almost total absence of such explanations exists among the interviews and observations in which I engaged there in two hundred hours of interviews and observations over two years. I took this utter lack of "happy talk" about television to be indicative of the community's activities.
13. See Bill Nichols, *Representing Reality* (Bloomington: Indiana University Press, 1991), 3.
14. Ibid., 31.
15. Roger Silverstone, *Television and Everyday Life* (London: Routledge, 1994).
16. Daniel Hallin, *We Keep America on Top of the World: Television Journalism and the Public Sphere* (London: Routledge, 1994), 102.
17. The Glasgow Media Group, *Really Bad News* (London: Writers and Readers Publishing Society, 1982), 143.

18. John Fiske, "Moments of Television: Neither the Text nor the Audience," in *Remote Control: Television, Audiences, and Cultural Power*, ed. Ellen Seiter et al. (London: Routledge, 1989), 56–78.
19. See John Fiske and John Hartley, *Reading Television* (London: Methuen, 1978), 112, 125.
20. Ibid., 125–126.
21. Concerns about age stereotyping distinguished many of the comments I heard about Andy Rooney. People felt that Rooney's drollery was unfairly representative of their age cohort and, by association, made them out to be buffoons. Several residents similarly criticized characters in the situation comedy *The Golden Girls* for the same reason.
22. Hallin, *We Keep America*, 100.
23. Richard Campbell, "Securing the Middle Ground: Reporter Formulas in 60 Minutes," in *Critical Perspectives on Media and Society*, ed. Richard K. Avery and David Eason (New York: Guilford Press, 1991), 65–293.
24. Sonia Livingstone and Peter Lunt, *Talk on Television: Audience Participation and Public Debate* (London: Routledge, 1994), 22.
25. Ibid., 57.
26. Ibid.
27. Haim Hazan, *The Limbo People: A Study of the Constitution of the Time Universe among the Aged* (London: Routledge and Kegan Paul, 1980).

Four Preaching to the Unseen Choir

1. See, for example, Richard H. Davis, *Television and the Aging Audience* (Los Angeles: University of Southern California Press, 1980).
2. The people whose stories are related here were identified through sources at the Milwaukee Access Telecommunications Authority when we told them we were looking for elders involved with production. Generally, we conducted one-and-one-half hour-long interviews with people in their homes or offices and followed up later with additional questions. We attended tapings of programs and studied videotapes of additional cablecasts.
3. See Laura A. Reese and Ronald E. Brown, "The Effects of Religious Messages on Racial Identity and System Blame among African-Americans," *Journal of Politics* 57, no. 1 (1995): 24–43, and Barbara P. Payne, "Religious Patterns and Participation of Older Adults: A Sociological Perspective," *Educational Gerontology* 14 (1988): 255–267.
4. Harold G. Koenig, *Research on Religion and Aging: An Annotated Bibliography* (Westport, Conn.: Greenwood Press, 1995), 152.
5. Jeffrey S. Levin, Robert Joseph Taylor, and Linda M. Chatters, "Race and Gender Differences in Religiosity among Older Adults: Findings from Four National Surveys," *Journal of Gerontology* 49, no. 3 (1994): 137–145.
6. Walter F. Pitts, *Old Ship of Zion: The Afro-Baptist Church in the African Diaspora* (New York: Oxford University Press, 1993).
7. Penny A. Ralston, "Learning Needs and Efforts of the Black Elderly," *International*

Journal of Aging and Human Development 17, no. 1 (1983): 75–88, and Will D. Morrison, "The Black Church as a Support System for Black Elderly," *Journal of Gerontological Social Work* 17, nos. 1–2 (1991): 105–120.

8. Peter J. Paris, *The Spirituality of African Peoples: The Search for a Common Moral Discourse* (Minneapolis: Fortress Press, 1995).
9. See Reese and Brown, "Effects of Religious Messages."
10. In the interest of consistency, all the names of the five research subjects have been changed for anonymity. Several of them made statements in the interviews that they preferred not to have connected to them directly by name.
11. See Levin et al., "Race and Gender Differences," S137–38.
12. See Rick Szykowny, "The Threat of Public Access," *Humanist* 54, no. 3 (May/June 1994): 15–22.
13. Ibid., 18.
14. Senior producers, who seem to reflect more traditional African American views, tend to be represented in the Christian programming rather than the Islamic.
15. Awkwardly, we obliged Yvonne's request. This was little problem for me, a Lutheran, and, luckily, the graduate student, a Unitarian, was a good sport.
16. William P. Nye, "Amazing Grace: Religion and Identity among Elderly Black Individuals," *International Journal of Aging and Human Development* 36, no. 2 (1992–1993): 103–114.
17. For an argument about the linkage between working-class respectability and religiosity, see Beverly Skeggs, *Becoming Respectable: An Ethnography of White Working-Class Women* (London: Sage, in press). Although Skeggs's work is concerned with whites, the argument extends to members of the black working class who involve themselves in churches, in part in an effort to attain the middle-class respectability (status) that might otherwise elude them.
18. Of course the shortcoming of our methodology was that our research subjects were able to shape their self-presentations for our consumption, but their interview styles seemed to match their television presentation styles.
19. Richard H. Davis, *Television and the Aging Audience* (Los Angeles: University of Southern California Press, 1980).
20. For the most incisive critique of the politics of interviewing in the ethnographic tradition, see the work of Ien Ang, for example, her recent book *Living Room Wars: Rethinking Media Audiences for a Postmodern World* (London: Routledge, 1996).
21. Jason Roberts, "Public Access: Fortifying the Electronic Soapbox," *Federal Communications Law Journal* 47, no. 1 (October 1994): 123–152; see 151.
22. Also, a study of Milwaukee residents published in 1988 observed that 65 percent of residents "never" watched public access and only a slight majority was aware of its existence. There is a paucity of research exploring the change in audience relationships to public access since, and we have no reason to believe that the statistics are substantially different today. See Gregory S. Porter and Mark J. Banks, "Cable Access as a Public Forum," *Journalism Quarterly* 65 (Spring 1988): 39–45.
23. Benjamin Barber, *Strong Democracy: Participatory Politics for a New Age* (Berkeley and Los Angeles: University of California Press, 1984). Barber's work is

discussed by Richard E. Sclove in *Democracy and Technology* (New York: Guilford Press, 1995).

24. Sclove, *Democracy and Technology*, 30.
25. Ibid., 79, 108.
26. Nancy Fraser, "Rethinking the Public Sphere: A Contribution to the Critique of Actually Existing Democracy," in *Habermas and the Public Sphere*, ed. Craig Calhoun (Cambridge: MIT Press, 1992), 109–142; quoted material is from 122–124.
27. Alexander Meiklejohn, *Political Freedom: The Constitutional Powers of the People* (New York: Oxford University Press, 1965), 26.
28. Theodore L. Glasser and Stephanie Craft, "Public Journalism and the Search for Democratic Ideals," paper, International Communication Association, Chicago, 1997.

Five A Pinch of Salt

1. Such a portrayal gets to the heart of popular culture's propensity for, in Edward Said's terms, "orientalizing" Asians. See Said's *Orientalism* (New York: Vintage Books, 1979).
2. See, for example, the work done among gerontologists on aged lesbians, such as Monica Kehoe, "Lesbians over 65: A Triply Visible Minority," *Journal of Homosexuality* 12, nos. 3–4 (1986): 139–152, and Charity V. Schoonmaker, "Aging Lesbians: Bearing the Burden of Triple Shame," *Women and Therapy* 16, no. 2 (1993): 21–31.
3. Abhik Roy and Jake Harwood, "Underrepresented, Positively Portrayed: Older Adults in Television Commercials," *Journal of Applied Communication Research* 25, no. 1 (1997): 39–56. See also Walter Gantz, H. M. Gartenberg, and C. K. Rainbow, "Approaching Invisibility: The Portrayal of the Elderly in Magazine Advertisements," *Journal of Communication* 30 (1980): 56–60, and Kathleen Woodward, *Aging and Its Discontents: Freud and Other Fictions* (Bloomington: Indiana University Press, 1991).
4. I do not mean to suggest that ethnic and racial minority group elders share the same economic benefits as whites. For an explanation of the inferior material living standards of minority elders, see James S. Jackson, Shirley A. Lockery, and F. Thomas Juster, "Minority Perspectives from the Health and Retirement Study," *Gerontologist* 36, no. 3 (1996): 282–284.
5. See Raymond M. Berger, *Gay and Gray: The Older Homosexual Man*, 2d ed. (New York: Hayworth Press, 1996).
6. The chapter represents conversations with elders either in their homes or at congregate meals sites in the Milwaukee area. I had no systematic method of identifying minority groups but wished to represent a range of voices. Contacts with nonprofit organizations that cater to the various groups mentioned here led to the recruitment of these volunteer participants. Generally, two-hour depth interviews were conducted with each person, except in the case of the Hmong elders, who were interviewed in groups of two with a translator; these interviews were briefer. A research assistant helped with some of the interviews.
7. Dutch cultural critic Jan Nederveen Pieterse has exhaustively detailed the ap-

propriation of African images by Europeans, Americans, and other Westerners. See his book *White on Black: Images of Africa and Blacks in Western Popular Culture* (New Haven: Yale University Press, 1992).

8. See Lyn McCredden, "Toward a Critical Solidarity: (Inter)Change in Australian Aboriginal Writing," in *Cross-Addressing: Resistance Literature and Cultural Borders*, ed. John C. Hawley (Albany: State University of New York Press, 1996), 13–34; see 15.
9. Jennifer Sabbioni, Kay Schaffer, and Sidonie Smith, eds., *Indigenous Australian Voices* (New Brunswick, N.J.: Rutgers University Press, 1998).
10. Kobena Mercer, *Welcome to the Jungle: New Positions in Black Cultural Studies* (London: Routledge, 1994).
11. Simon Jones, *Black Culture, White Youth: The Reggae Tradition from JA to UK* (London: Macmillan Education, 1988).
12. John Fiske has written about the white appropriation of rap, explaining that white audiences who "hear" rap do not necessarily "listen" to it with a sophisticated understanding of black discourse. While whites tend to take rap lyrics literally, Fiske suggests, blacks maintain an informed distinction between *narratives about* violence and actual violence *itself*. See *Media Matters: Race and Gender in U.S. Politics* (Minneapolis: University of Minnesota Press, 1996).
13. In staking out a position for the personalized "place" of television in everyday life, Silverstone responds to the influential essay in which Margaret Morse suggests television viewing as a very impersonal experience wherein the viewer takes up an unspecific, unattached, liminal state akin to freeway driving or mall roaming. Silverstone wants to anchor the viewer—in a setting as well as within a set of perspectives—much more than Morse does. In probing the tension between the idea of the "nomadic" subject and the "free" subject, Silverstone sympathizes with the position of Janice Radway rather than the more optimistic John Fiske. See Margaret Morse, "An Ontology of Everyday Distraction: The Freeway, the Mall, and Television," in *Logics of Television: Essays in Cultural Criticism*, ed. Patricia Mellencamp (Bloomington: Indiana University Press, 1990), 193–221; Janice Radway, *Reading the Romance: Women, Patriarchy, and Popular Literature* (Chapel Hill: University of North Carolina Press, 1984) and "Reception Study: Ethnography and the Problems of Dispersed Audiences and Nomadic Subjects," *Cultural Studies* 2, no. 3 (1988): 359–376; and John Fiske, "Moments of Television: Neither the Text nor the Audience," in *Remote Control: Television, Audiences, and Cultural Power*, ed. Ellen Seiter et al. (London: Routledge, 1989), 56–68.
14. David Morley, *The "Nationwide" Audience* (London: British Film Institute, 1980).
15. Ironically, the methodological tool chest of the social scientist provides readier access to conveying the complicated terms of people's social makeup than does the comparatively plodding narrative of cultural studies. In other words, it is efficient to tabulate people as amalgams of variables ("Middle-class black elders do this while middle-class white elders do that") than it is to convey distinctions and blurred boundaries through the rich interpretations that are the hallmark of cultural studies. But, as painting these distinctions and overlaps is part of our basic project, we approach them as bits of a pointalist picture.

16. Ien Ang, *Desperately Seeking the Audience* (London: Routledge, 1991), and Radway, "Reception Study."
17. Jenny Hockey and Allison James, *Growing Up and Growing Old: Ageing and Dependency in the Life Course* (London: Sage, 1993), 46.
18. Gerontologists Ronald J. Angel and Jacqueline L. Angel have detailed the difficulties that minority elders face in late modern life in their book *Who Will Care for Us?: Aging and Long-Term Care in Multicultural America* (New York: New York University Press, 1997).
19. See, for example, Nancy Wood Bliese, "Media in the Rocking Chair: Media Uses and Functions among the Elderly," in *Inter/Media: Inter-personal Communication in a Media World*, ed. Gary Gumpert and Robert Cathcart (New York: Oxford University Press, 1986), 573–582.
20. For an explanation of intergenerational difficulties among gay men, see Berger, *Gay and Gray*.
21. For support of Ned's observation, see Alexander Doty, *Making Things Perfectly Queer: Interpreting Mass Culture* (Minneapolis: University of Minnesota Press, 1993).
22. Feminist television scholars have with some consistency found that women report watching television amid the performance of domestic duties. See, for example, Ellen Seiter, Hans Borchers, Gabrielle Kreutzner, and Eva-Marie Warth, "Don't Treat Us Like We're So Stupid and Naïve: Toward an Ethnography of Soap Opera Viewers," in Seiter et al., *Remote Control*, 223–247.
23. In these ways, Mildred's experience with television was resonant of the situation I found at Woodglen.
24. As with many of the women I talked about in chapter 2, *Murder, She Wrote* drew Cynthia because of the appeal of an independent woman in the driver's seat. Although she liked other mysteries because she enjoyed seeing the puzzle solved, the Lansbury character was important to her as an aging lesbian, constituting proof that older women did not need men to thrive.
25. The use of American broadcasting for the acquisition of English skills has a long history, and it is commonplace for English as a Second Language classes to rely on television as a "homework" tool.
26. Milwaukee is an extremely segregated city, where whites cluster in sections and nonwhites in others, a function shaped to a high degree by economics. Some nonwhite neighborhoods are almost entirely African American, some mostly Hispanic. Native Americans cluster in a few pockets of the city with the largest population being on the multiethnic south side.
27. Red Cloud liked one particular local news show, because she said one of the announcers was an Indian, but she was unsure of his name.
28. In his study of homeless Native American men, John Fiske observed this critical tendency in noting that the men derived keen pleasure from watching the *part* of a Western in which the Indians appeared victorious—overtaking the wagon train, for instance—and then turned the set off before the narrative inevitably proceeded to the white defeat of the Indians. See Fiske, *Understanding Popular Culture* (Boston: Unwin Hyman, 1989).

29. I interviewed Frankie shortly after the bombing at the U.S. Olympic Center in Atlanta.
30. See the uses and gratifications work of Richard H. Davis, *Television and the Aging Audience* (Los Angeles: University of Southern California Press, 1980).
31. Stuart Hall, "Encoding/Decoding," in *Culture, Media, Language*, ed. Stuart Hall, Dorothy Hobson, Andrew Lowe, and Paul Willis (London: Hutchinson, 1980).
32. As John Fiske has put it, "Social identities and thus social alliances always include elements of class, race, and gender, but the proportions of the mix cannot be predicted or generalized." See *Media Matters*, 110.
33. James Clifford, "Traveling Cultures," in *Cultural Studies*, ed. Lawrence Grossberg, Cary Nelson, and Paula Treichler (New York: Routledge, 1992), 96–112.
34. Marie-Louise Mares and Joanne Cantor, "Elderly Viewers' Responses to Televised Portrayals of Old Age: Empathy and Mood Management versus Social Comparison," *Communication Research* 19, no. 4 (1992): 459–478.
35. Pierre Bourdieu, *In Other Words* (Cambridge, England: Polity Press, 1990). Jim McGuigan has explored this subject in *Cultural Populism* (London: Routledge, 1992).

Six Theorizing Television in Old Age

1. Ien Ang, *Living Room Wars: Rethinking Media Audiences for a Postmodern World* (London: Routledge, 1996), and Lynne Joyrich, *Re-Viewing Reception: Television, Gender, and Post-Modern Culture* (Bloomington: Indiana University Press, 1996).
2. See, for example, Hermann Bausinger, "Media, Technology, and Daily Life," *Media, Culture, and Society* 6 (1984): 343–351, and Roger Silverstone, *Television and Everyday Life* (London: Sage, 1994).
3. Janice Radway, "Reception Study: Ethnography and the Problems of Dispersed Audiences and Nomadic Subjects," *Cultural Studies* 2, no. 3 (1988): 359–376.
4. Wilbur Schramm, "Aging in Mass Communication," in *Aging and Society*, vol. 2, *Aging and the Professions*, ed. M. W. Riley, J. W. Riley Jr., and M. E. Johnson (New York: Russell Sage Foundation, 1969), 352–376.
5. I am thinking here of such statements as this one by Richard H. Davis and G. Jay Westbrook: "[The elderly] are heavy viewers, with older females spending more time with television than any other audience. . . . [The elderly] are 'embracers,' accepting with little judgment what is placed before them." See Davis and Westbrook, "Television in the Lives of the Elderly: Attitudes and Opinions," *Journal of Broadcasting and Electronic Media* 29, no. 2 (1985): 209–214. I am also thinking of the television industry's traditional programming strategy of ignoring older viewers as a market force because they assume that elders can be depended on both to watch and to absorb content while the networks have to work to hook younger viewers for advertisers. This argument is well documented and is highlighted in chapter 2, but for more reading, see Paul Farhi, "On TV, Madison Ave. Sets the Dial for Youth," *Washington Post*, September 16, 1994, A1, 21.
6. M. Powell Lawton, Miriam Moss, and Louise M. Duhamel, "The Quality of Daily

Life among Elderly Care Receivers," *Journal of Applied Gerontology* 14, no. 2 (1995): 150–171; see 163.

7. Betty Friedan, *The Fountain of Age* (New York: Simon & Schuster, 1993).
8. I would not pretend to address competently the television viewing experience of the Alzheimer's patient. I have been careful to avoid interviewing such persons for the purpose of this book. That is a topic for another project.
9. See, for example, Nancy Wood Bliese, "Media in the Rocking Chair: Media Uses and Functions among the Elderly," in *Inter/Media: Inter-personal Communication in a Media World*, ed. Gary Gumpert and Robert Cathcart (New York: Oxford University Press, 1986), 573–582, and R. W. Kubey, "Television and Aging: Past, Present, and Future," *Gerontologist* 20 (1980): 16–35.
10. Times Mirror Center for the People and the Press, *The Role of Technology in American Life: Technology in the American Household* (New York: Times Mirror Co., 1994).
11. See J. Walker Smith and Ann Clurman, *Rocking the Ages: The Yankelovich Report on Generational Marketing* (New York: HarperBusiness, 1997).
12. British sociologist Andrew Blaikie, following the work of postmodernist Mike Featherstone and others, frets that the lengthening "active midlife plateau" will cause society to blot out from its consciousness "the darker side of dying." See Blaikie, "Images of Age: A Reflexive Process," *Applied Ergonomics* 24, no. 1 (1993): 51–51; see 55.

Index

About the Author

Karen E. Riggs is an assistant professor in the Department of Mass Communications at the University of Wisconsin–Milwaukee. She has published articles on television and sports and on television and the elderly in sociology and mass communication journals.